ROLLS - ROYCE HERITAGE TRUST

HENRY ROYCE
— mechanic

Donald Bastow

HISTORICAL SERIES No 12

Published in 1989 by the
Rolls-Royce Heritage Trust
PO Box 31 Derby England

ISBN: 0-9511710-4-6

The Historical Series is published as a joint initiative by the
Rolls-Royce Heritage Trust and the Sir Henry Royce Memorial Foundation.

Also published in the series:

No. 1 Rolls-Royce—the formative years, 1906-1939
Alec Harvey-Bailey, published by RRHT

No. 2 The Merlin in Perspective—the combat years
Alec Harvey-Bailey, published by RRHT

No. 3 Rolls-Royce—the Pursuit of Excellence
Alec Harvey-Bailey and Mike Evans, published by HRMF

No. 4 In the Beginning—the Manchester origins of Rolls-Royce
Mike Evans, published by RRHT

No. 5 Rolls-Royce—the Derby Bentleys
Alec Harvey-Bailey, published by HRMF

No. 6 The Early Days of Rolls-Royce—and the Montagu family
Lord Montagu of Beaulieu, published by RRHT

No. 7 Rolls-Royce—Hives, the Quiet Tiger
Alec Harvey-Bailey, published by HRMF

No. 8 Rolls-Royce—Twenty to Wraith
Alec Harvey Bailey, published by HRMF

No. 9 Rolls-Royce and the Mustang
David Birch, published by RRHT

No. 10 From Gipsy to Gem—with diversions
Peter Stokes, published by RRHT

No. 11 Armstrong Siddeley—the Parkside story, 1896-1939
Ray Cook, published by RRHT

Cover : Sir Henry Royce—a well known portrait photograph

Printed by Bemrose Security Printing, Derby

CONTENTS

FOREWORD

Here is an account of the life and work of Sir Henry Royce which is unique. Donald Bastow has written of the West Wittering days, not as an historian, but from his personal memories, for he was a member of that elite cadre, Sir Henry's personal design team. That was the first of his many qualifications for the authorship of this volume and I have to say "was", for Donald sadly died shortly before this volume went to press.

Donald took his BSc (Eng) with first class honours at University College, London, gaining the Archibald Head medal for the best mechanical engineering student of the year. He then joined the Daimler Company Limited and gained experience of the motor car industry over a three year post graduate traineeship. Leaving Daimler he joined Rolls-Royce as a designer and it was in the last years of Sir Henry's life that he worked with him at West Wittering.

When Sir Henry died in 1933, Donald, together with his colleagues, moved to Derby to continue designing. He became deeply involved in the perfection of independent front suspension, which was introduced initially at Derby on the Phantom III, and then on the Wraith and Bentley Mk V in the closing stages of the 1930s. During the war the Chassis Division engineering team moved to Belper and there Donald continued with the design of suspension systems, but on the Cromwell tank rather than motor cars. He headed the design team on this work.

When the war was over, Donald left Rolls-Royce for Lagonda Motors Ltd. His first year there was as Technical Production Engineer, but this was followed by being appointed assistant to the Technical Director - W.O. Bentley. For three years he worked with W.O. on the design and development of the post war 2.6 litre Lagonda. The David Brown organisation then bought the Lagonda and for a further three years Donald remained with W.O. in his assignment as consulting engineer to Armstrong Siddeley Car Division. They designed a car for Armstrong Siddeley which got as far as the manufacture of prototypes.

Donald's six years with W.O. were followed by a short period with BSA as Manager, Engine and Mechanisms Laboratory, before he was appointed Chief Engineer of Jowett Cars Limited. There he was responsible for continued development of the Javelin, as well as for the design and development of a new range of two and four cylinder cars and vans, including the new Jupiter sports car.

When Jowett was closed Donald moved for a year to Metalastik and then to Coventry Climax Engines Ltd where he became Chief Engineer, Electric Vehicles. His next post was as Design Director of Birfield Engineering Limited, later Chief Engineer of Hardy Spicer Limited in addition. Further design appointments followed with Dobson and Barlow Limited and TMM (Research) Limited before he started his own consultancy business, in which he continued until his death, specialising particularly in the automotive field.

Donald was much involved in the activities of the Institution of Mechanical Engineers of which he was a Fellow, and had presented many papers over the years on motor car specialisms. He had served as Member of Council and Chairman of the Automobile Division. In 1984 he was elected a Fellow of America's Society of Automotive Engineers and he was also a Member of the French Société des Ingénieurs de l'Automobile. Furthermore he served on many bodies including the Vehicle Committee of the Road Research Board and Council of the Motor Industries Research Association.

This is not Donald's first book. He authored "W.O. Bentley - Engineer", published by Haynes in 1978, and "Car Suspension and Handling", published in the following year by Pentech Press. He was also a member of the founding committee of the Technical Writers Group of the Society of Authors. We are indeed privileged to publish this volume. It gives me great personal pleasure to see Donald's story in print for I was fortunate to have been able to count Donald as a friend over a great many years. He will be greatly missed by all who had the privilege of knowing him.

MH Evans
Chairman Rolls-Royce Heritage Trust

AUTHOR'S PREFACE

A talk about the writing of biographies which I attended some time ago contained the suggestion that the final work would incorporate only about 15% of the material which the compiler had amassed. In this booklet the jettisoning has had to be even more severe: on the basis of relative word contents it has been possible to retain only a little over 2% instead of the 15% referred to above. Luckily the position is not quite so bad as that. If one were to accept the dictum, referring to technical writing, that a good illustration is worth a thousand words, the abundance of illustrations included, making some allowance for exaggeration, has partly made up for the loss.

The intention here however has not been to provide a comprehensive biography, but enough to show how a considerable organisation owes its present existence and scale to the application of the abilities and aspirations of one man in the engineering of its products. In the other aspects of a company's activities this one man needed to be, and was, ably backed by devoted and almost equally determined individuals.

The choice of material to be published here has been governed by several factors: the contents of the existing booklets in this series; new information on the Royce family; establishing the reasons for the poverty and hardship of Royce's early life; demonstrating the logicality of the various developments in the products; and a desire to show some more detail on the results of the isolation of Royce from the Derby works on himself and his immediate staff. One effect of that isolation was the need to keep abreast of developments, in the car field in particular, by reading. I believe we had copies at West Wittering of 'Autocar', 'Motor', the SAE Journal, 'Automotive Industries', 'La Vie Automobile' and publications from the French Société des Ingénieurs de l'Automobile; we certainly had some of them, and all of them at Derby later. We would get our own copies of the publications of the Institution of Automobile Engineers. There was no local centre of this Institution and therefore no meetings we could attend. I need hardly say that we were expected to study such publications in our spare time!

Frederick Henry Royce was a remarkable man: in his prime he was boisterously and persistently active, intolerant, parsimonious, in some matters even cheeseparing. He acquired a remarkable grasp of the technical needs of two quite diverse industries. By his own omissions he eventually lost his boisterousness, but by then the company had acquired the necessary momentum to maintain its progress.

I have needed and obtained a great deal of help from various sources. I would like to say how much this has been appreciated. First and foremost, I have had from Michael Evans enduring help, advice, suggestions and encouragement. Others who have made very useful contributions in different ways are Harry Biraben, Tom Clarke, John Fasal, Fred Hardy, Alec Harvey-Bailey, Bill Morton, Mrs Beryl Neish (nee Royce), Rolls-Royce plc, Rolls-Royce Motor Cars Ltd., Paul Tritton and John Welch. I have also abstracted information from many previously published sources and would like to acknowledge my considerable indebtedness to their authors and the research they must have put into their books and articles.

LIST OF ROLLS-ROYCE PERSONAL REFERENCES

(IN ORDER OF THEIR MENTION)

Sir Henry Royce	R
Ernest A Claremont	EAC
The Hon Charles S Rolls	CSR
Claude Johnson	CJ
A J Rowledge	Rg
Ernest W Hives	Hs
R W Harvey-Bailey	By
Maurice Olley	Oy
Albert G Elliott	E
Bernard I Day	Da
Arthur Wormald	Wor
Bill Clough	Cgh
W A 'Roy' Robotham	Rm
A Cyril Lovesey	Lov
James E Ellor	Lr
William 'Bill' G Hardy	Hdy
H Ivan Evernden	Ev
Thomas Nadin	Na
Charles L Jenner	Jnr
Samuel F Pottinger	SFP
Don Eyre	DE
A E 'Tony' Cooke	AEC
Harry E Biraben	HEB
John DeLooze	D
Basil Johnson	BJ
Fred Hardy	FJH
Donald Bastow	DB

HENRY ROYCE - MECHANIC

1. Introduction

The title of this study derives from Royce's own description of himself as a mechanic. The Royce family background, and an outline of his life and achievements, are felt to be necessary to understand his character. His life and works are dealt with much more fully elsewhere. In dealing with these therefore brevity has been the keyword. To save time generally, and to conform to Rolls-Royce Heritage Trust practice, the man who became Sir Henry Royce will be referred to here by his Rolls-Royce symbol 'R'. Their equivalent symbols are also used for his colleagues if they crop up often.

2. The Royce Family

R believed that he was descended from a Welsh master bowman called Rhys or Ryce, one of the followers who helped Henry Tudor to the English throne in 1485 at the battle of Bosworth Field, and was duly rewarded. The family has certainly, and more recently, been traced back to the time of Good Queen Bess (1). R's father was descended from generations of farmers and millers. R's grandfather, Henry William Royce (2), married three times and was a pioneer in the installation of steam power in water mills. Two cousins, the Cooper brothers William and George, emigrated to Toronto in Canada in 1826. William died from blood poisoning as the result of a scratch from a bear kept to entertain customers at his Toronto public house, but George bought the 400 acre plot they had planned to get jointly (3, 4) and later married his brother's widow. They had no children so R's uncle Allan (1835-1902) was invited to join them, grew up with them and later became George's manager and eventually his heir. He married Sarah Jane Gilbert, George's wife's niece. Hard and soft wood timber from the plot was felled and sold and a farm established. The Canadian families prospered.

Henry William Royce too was invited to join cousin George Cooper in Toronto about 1870, together with his third wife Mahala (Lane) and most of his family. Two exceptions were R's father James, by that time seriously ill, and an uncle John Charles who preferred to remain behind in England. He was Henry William's first son by his third wife.

A member of another branch of the family, Robert Royce, had emigrated to Connecticut in the reign of Charles II. Two of his sons, Nathaniel and Nehemiah, were connected with Wallingford in Connecticut from its inception. Nehemiah, having married a member of the later famous Morgan banking family, built a house in Wallingford which was known afterwards as the "Washington Elm House". There grew in front of it a huge elm under which George Washington said goodbye to the citizens of Wallingford after a visit there in 1775. The house, thanks to the efforts of two of Nehemiah's descendants, was saved from demolition in 1924. It was moved to a site nearby and eventually presented to the Choate School Foundation in 1962 to serve as a museum.

3. **Early Days**

James (1831-1872), in the family tradition, had a farming training until he moved on to milling, renting the mill at Castor, Northamptonshire, in 1852. He had just married Mary King, daughter of a progressive large scale farmer. They and their three daughters and first son moved a mile or so across the river Nene to the mill at Alwalton in Huntingdonshire in 1858. R (1863-1933) was born on 27th March at this mill, rented from the Dean and Chapter of Peterborough. It had steam as well as water power, almost certainly installed by James, following in his pioneer father's footsteps. James was bright but unreliable, and lacked application. This may have been due to his having Hodgkin's disease, which induces anaemia and eventually intermittent fever. He was in money trouble as early as 1863, when he mortgaged the Alwalton mill lease to the London Flour Co. The trouble got worse and he went to London to work for this company in 1867, taking his two sons but leaving his wife, daughters and debts behind. He died in 1872 at the age of 41 in a poor house, an exception in a prosperous family of farmers and millers.

It is not surprising that R's early years of poverty and hardship left lasting and unpleasant memories. Even before he was four he had been birdscaring, for money, in the fields near Alwalton. Recollections of early days in London are of cold and hunger; on one evening when he was left alone in a cold house he found it warmer to join the dog in his kennel than to stay indoors. As James' condition deteriorated R's mother found housekeeping jobs nearer to them but could not have R with her and he had to remain in lodgings. After his father died he sold newspapers for W.H.Smith. A year at school was followed by a job delivering telegrams in Mayfair. At the age of fourteen an aunt on his mother's side of the family at Fletton on the outskirts of Peterborough, with whom R had spent some holidays, agreed to pay the £20 a year for him to be an apprentice at the Great Northern Railway works in Peterborough. He lodged with Mr and Mrs Yarrow. In his spare time he went to evening classes in English and mathematics, learned a good deal more of machining and fitting in Mr Yarrow's garden workshop and sold more newspapers. Some articles about him after his death called him a "terror for work"! He was fanatically determined to make up for his earlier lack of formal education and make the most of the opportunity that had been presented to him.

After three years at Peterborough the aunt was unable to continue her support. Within a fortnight R found work in Leeds, with Greenwood and Batley, as a toolmaker, making machines for the Italian Arsenal then being built. He was paid 55p for a 54 hour week. A spare time interest in electricity and its applications then led to a job with the Electric Light and Power Co. in London. At evening classes he favourably impressed Professor William Ayrton. At work his help in some experiments also impressed Hiram Maxim, whose electric light bulb patents the company had acquired. Altogether he did well enough to be sent as first electrician to the associated company, Lancashire Maxim and Weston Electric Co., doing theatre and street lighting in Liverpool. A complete lighting installation for several streets, for which he had the technical responsibility, was accepted by the Corporation in March 1884 just before R's 21st birthday. By the end of May the company had failed. R had saved £20; a friend, E.A.Claremont, also with electrical training, contributed £50 and together they formed F.H. Royce & Co., electrical engineers, and rented premises at Cooke Street, Manchester. They started in No.1a. The eventual development there was considerable (5, 6).

Gas was still the main source of light, public and private. Swan had made his first successful carbon filament electric light bulb only five years before. Although there had been rapid development it had made little impact on public lighting, for which carbon arc lights, three times as efficient optically, were naturally preferred. There were very few public supplies of electricity. If any organisation wanted electricity for any purpose it would almost certainly have to install and operate its own generating plant. F.H.Royce & Co. started with small items for individual sales, such as an electric bell set, powered by Leclanché cells. The fact that items such as bulb holders, switches, fuses, filaments and probably complete bulbs, and registering instruments, were also early products suggests a fairly early entry into the sub-contract business. They moved on to complete installations. Profits fed back led to dynamos, motors and eventually winches and cranes. R found how to get sparkless commutation and made other improvements. He was the technical partner, completely engrossed in his work, often neglecting to break off either to eat or even to sleep. Claremont (EAC) dealt with the sales and business aspects, and unlike his partner had a fetish for physical fitness. The timing clock (7) and the dynamos or motors (8, 9, 10, 11) relate to this period.

The photograph of R (12) in 1896 and of EAC (13) show the two partners. An early workshop scene is interesting (14).

The business grew enough to allow the partners to marry in 1893 the two daughers of the late Mr Alfred Punt, a London licensed victualler and printer. In June 1894 it became a limited company. It continued to grow. Orders in hand in October 1897 were £6,000; by the following March £9,000; and eleven months later £20,000, in spite of growing competition. Later in 1899 the share capital was increased by £30,000 to allow an extra works to be built at Trafford Park; Cooke Street had become too small.

The new establishment and its relationship to the W.T.Glover cable making firm are seen from air and ground level (15, 16). The Trafford Park works had room to manufacture gantry cranes (17, 18); one is still in use at the Rolls-Royce Derby works (19, 20, 21).

The Royces lived at first in a rented house in Charlton-cum-Hardy; he later built a house in Legh Road, Knutsford, which of course had its own electrical installation (22). The plaque in the gable end (23) has the reversed initials FH for Frederick Henry and MG for Minnie Grace. Calling the house Brae Cottage seems in keeping with R calling himself a mechanic. The house was designed by Waterhouse, a leading architect of the day, who was responsible also for the Manchester Town Hall and many other public buildings.

R added gardening to his other activities, often carried out by electric light because of the time he spent at work. His wife, Minnie Grace, ran the household, ordered his seeds, generally looked after and admired him. He had brought his mother, now retired, to be in lodgings near him and usually looked in on her on his way home. He also bought his first motor vehicle, a De Dion quadricycle, and his education in that sphere began.

In 1902 R collapsed as a result of continuous overwork. He was too single-minded to eat during the day unless strongly prompted. Overseeing the building and layout of the Trafford Park works had taken an extra toll; and business had suffered from cheap foreign competition and the Boer war. Even so R didn't want to take a break. His wife Minnie Grace however had relatives in South Africa and decided she wanted to visit them. After ten weeks away R was mentally and physically refreshed. He was then induced by the combination of Dr. Campbell-Thompson and EAC to replace the quadricycle by a car. It seems likely that R himself chose the model, a 10hp twin cylinder Decauville (24, 25), made by a French firm of light locomotive and car builders, because of the reputation for reliability which the firm had gained for its products by observed trials (26, 27, 28, 29, 30, 31). A torque arm on the rear axle (32) does not appear on the car beautifully restored by John Welch and is not shown in the arrangement drawings (33, 34, 35).

4. Cars Before World War 1

R was dissatisfied with both the quality and reliability of his Decauville and tried various improvements. One of these was the mechanical operation of the overhead inlet valves (36). From experience with this car and those of friends and neighbours R decided he could do better himself and in 1903 got Board permission to build three cars of his own design at Cooke Street, in spite of EAC's opposition to the idea. Henry Edmunds (37), a fellow director and experienced automobilist, encouraged R in this. Making three cars implied a serious intention of adding to the product range. The addition of "Royce Petrol Motor Cars" to the list of products on the firm's headed paper confirms this intention, even before the fateful meeting between R and the Hon. C.S. Rolls (CSR), referred to below, took place. Royce Ltd. needed new products to compensate for business lost as a result of German and American competition in the electrical field, selling cheaper but inferior products. For a man of forty to make such a change in direction seems bold but R had proved in the last nineteen years that from unpromising beginnings, little formal education and abbreviated training he could establish himself in the growing electrical industry. It is generally easier to develop with a new industry than to break into an established one. R minimised the risk and the development period by using the Decauville as a basis. He had it taken to pieces and every part drawn. The chassis of the first Royce 10hp twin was very similar to that of the Decauville in layout, wheelbase and track and had semi-elliptic springs front and rear (38, 39, 40, 41, 42). Differences lay in separate engine and gearbox units isolated from the effects of frame distortion, positively opened inlet valves instead of automatic, a long torque arm fixed to the back axle casing, the use of a single lever quadrant change for the three speed gearbox in place of two separate levers, steel on bronze bearings in place of the unforgiving steel on steel, the elimination of the seven gallon water tank by the use of a better radiator and the use of a lighter and cheaper audible warning than the 45lb bell. A complete car resembles the Decauville (43).

The first engine ran, for six hours, on 16th September 1903 (44). The first complete car ran on 1st April 1904. Within three weeks Edmunds had taken it on the "Side Slip Trials" organised by the (not yet Royal) Automobile Club, on which the first day's run was 145.5 miles, covered at a 16.5 mph average. It was exceptionally quiet and reliable from the beginning.

The Hon. C.S.Rolls (CSR) (45) was then selling Minerva and Panhard cars and wanted a quality British car, preferably with at least three cylinders, to supplement or replace them. Henry Edmunds, already mentioned, was a keen motorist, a committee member of the Automobile Club and a close friend of CSR and his manager Claude Johnson (CJ). He had sent CSR photos and plans before the first car ran. He borrowed the car for the "Side Slip Trials" to demonstrate the "Parsons" chains, made by a company of which he was chairman. CSR must have seen the car during these trials, in which it performed well, and friends of his had been driven in it. This made it easier for Edmunds to persuade CSR to visit Manchester with him, meet R over lunch at the Midland Hotel and try the car for himself. CSR was impressed by car and designer; R liked CSR. An agreement was reached verbally for CSR to sell all the cars R could make, and for them to be called Rolls-Royces (46, 47).

However good the 10hp car might be, it did not really fill CSR's needs for his main customers. So as a first step a decision was taken to make a three cylinder 15hp version using as many common parts as possible; this would be in line with CSR's previous ideas. The 15hp car represented more of a step forward and upward than might at first appear; the wheelbase was increased from 75″ to 103″ and the track from 48″ to 52″. The frame width however was retained at 36″ and this persisted, with three minor exceptions, for very many years. The car also had platform rear springing, with the rear ends of the rear half elliptic springs connected by crossed-axis shackles to the ends of an inverted semi-elliptic across the back of the car (48). Once the decision to produce the 15hp was taken (49), it became for R an obvious step to expand the range even further upmarket, with 20hp four (50) and 30hp six cylinder models (51). The track was expanded a further 4″ to 56″ for these models. Wheelbases were 114″ for the 'Heavy' 20hp, 116.5″ and 118″ for the 30hp. No wonder CSR called R the "greatest motor engineer in the world"! R represented for him the source of British quality cars for which he had been looking. For R too the arrangement was a godsend. Not only did it provide an opportunity for him to follow his predilection for quality products, it relieved him of any worries about selling these products and more than justified his decision to branch out into a new field.

Another result of this first meeting was a decision to change the shape of the radiator top tank to what has become an enduring characteristic of the marque (46, 47, 49, 50, 51). Credit has been given to Greek temples for the new shape; another explanation given to me by Ivan Evernden (Ev) was that the hot water from the top of the cylinders was led to the centre of the top tank and had to be spread across the top of the radiator block to travel down it. Because there would be a gradually decreasing amount of water to be accommodated as it moved towards the sides of the radiator it would be logical to reduce progressively the cross section of the tank, both to maintain a constant velocity of the water and to use the minimum amount of material. More often than not, actions are taken for a number of reasons. The new radiator shape represents a new, distinctive and easily distinguishable shape; it was also technically correct.

This began for R a truly remarkable three year period. Production 10hp engines had a centre main bearing to reduce the deflections of the rather long crank front end (52, 53). The three cylinder had separate cylinders, which crowded out the inlet valves to a position over the exhaust valves instead of over the bore (54, 55), the four and six cylinders had groups of the original twin cylinder block. For all these larger engines the crankshaft bearing diameter was increased from 1.25″ to 1.5″. The three cylinder had balance weights opposite each throw (56). A revised auxiliary drive layout, used on the 15hp and larger engines, in conjunction with a single camshaft (57), simplified the drive and considerably shortened the front end of the engine (44, 61).

A carburetter was devised to overcome the inherent enriching with speed of the usual jet in a single choke (58). It avoided the constant fiddling with the extra air lever which was otherwise necessary. The control system associated with the Royce carburetter allowed the governor to be overridden (59).

At the Paris Salon in December 1904 examples of two, three and four cylinder cars were shown, and a six cylinder engine (60, 61). The gearbox was improved. The platform rear springing introduced on the 15hp was continued on the larger cars. The 'Olympia' Show of February 1905 included a complete 30hp car (51). CSR knew the value of sporting successes and entered two 20hp cars for the September 1905 Isle of Man T.T. races. Weight and petrol consumption limits were imposed. Two specially lightened cars were prepared with overdrive "sprinting" gears added to the gearbox (62, 63). CSR had gearbox trouble on the first lap; Percy Northey on the second car finished a close second in the race. For the 1906 T.T., which CSR won, the cars had been further lightened. Wire wheels had by then become available and CSR asked why the cars had not been fitted with them. Because they were stronger sideways they allowed higher cornering speeds, and they were also lighter. R said they could not afford them; CSR said he must have them and would pay for them himself. Percy Northey again drove the second car but this suffered a broken front spring and failed to finish (64, 65, 66). The works load to which R and the whole of the Cooke Street establishment were being subjected at this time was considerable. R would have the burden of all the major technical decisions. The drawing office had to cope with a host of new components for the larger models: the single cylinder block for the three cylinder; gearboxes to take the higher torques and to include, initially only for the T.T. cars, the extra, overdrive, ratio; rear axles also to deal with the higher torques; and of course the crankcases and crankshafts. The shortened front end of the crankshaft, with its auxiliary drives already mentioned, reduced the extra space required to accommodate the longer engines. It also reduced the tendency to wobble of the front end of the crank which R had noticed and which led to his adding the centre main bearing to the twin. In this it was helped by putting up crank bearing sizes of the larger engines to 1.5″ from the 1.25″ of the twin. In keeping with the three bearing twin crankshaft the larger engines had four, five and seven bearing crankshafts. The complete lack of balance weights on the Decauville crankshaft, not even the rotating weights being dealt with, must have horrified R with his background of smooth running rotating machinery. He was quick to appreciate the simplicity and economy of putting the still large balance weights on the end webs of the twin (52). On the three, with three separate cranks at 120°, he had put two balance weights opposite each crank. The minimum amount of counterweight would be for the rotating parts only; this would leave out-of-balance couples in the plane of the cylinders. Extra counterbalancing, of a part of the reciprocating weights, (piston and upper part of each connecting rod) would reduce the maximum amount of the unbalanced couples but some of them would be across the engine. This would be an overall improvement. The four cylinder crank form was well established, two 180° twins back to back; primary forces and couples were balanced out, leaving only secondary forces unbalanced.

There was little precedent for the six cylinder crank arrangement; it is generally accepted that the Spyker of 1902 was the first ever six. The first production six had only appeared early in 1904. Because a six would probably be three twins in a row, there was a tendency to retain the 180° angle between the cranks in each twin, giving what was known as the 2+2+2 crank arrangement, the angle between successive twin cranks being chosen to get the most even firing intervals of 120° - 120° - 180° - 120° - 120° - 60°. There are two possible crank arrangements, successive twin cranks having either 60° or 120° between them. The 60° angle is superficially more attractive because it puts the foremost throw of the succeeding crank as nearly opposite to the last crank pin of its predecessor as possible. A more careful investigation, described in the Appendix for the more technically minded, shows it to have larger out-of-balance couples than the 120° angle variety, which does not require or benefit from balance weights. The unacceptable roughness of the original 30hp six suggests that R may have adopted the 60° angle. He retained the 180° twin balance weights, making six in all. Bill Morton, who saw one of these cranks, remembers seeing a lot of balance weights.

The roughness to which R objected seems likely to have been due to the out-of-balance couples, considerably larger than those of the three cylinder; he also noticed that the front end timing gears rattled. Some of this may have been due to the uneven firing intervals. R's first reaction to the rattle was to add a small front end flywheel. The major cause of the rattle was torsional vibrations of the crank, due to the torsional frequency of the crank assembly being lowered by the existence of the balance weights and the length of the crank and so bringing the third major vibration close to or within the running range. The front end flywheel, further reducing the frequency, resulted in broken crankshafts. In an experiment to find the best compromises on front flywheel inertia, R for quick results used a wooden adaptor between the flywheel and the crank. With this the vibration was less than when the wooden adaptor was replaced by a steel one; R also noticed that the clamped surfaces of the wooden adaptor showed scorch marks, implying relative movement. R had discovered by chance the principle of the Lanchester crankshaft torsional damper, later patented by Dr. F.W.Lanchester, of a front flywheel with friction damping on any relative movement between it and the crankshaft. R however did not immediately make use of this discovery; the solution of the vibration due to the unbalanced couples was staring him in the face: all he had to do was to adopt the principle used in the four cylinder crank, using two three cylinder cranks in mirror image form about the middle of the engine. With this arrangement all out-of-balance forces and couples are balanced out without the need for balance weights, and there are even firing intervals. That R realised this very quickly is shown by the picture of Albert Wheatman machining a six cylinder crank (67), with samples of the other cranks visible in the foreground. The absence of balance weights is confirmed by arrangement drawing (61), which also shows that R stuck to his, very light, front end flywheel. With this crank arrangement the only function of balance weights is to reduce main bearing loading, in particular on the centre main bearing of a seven bearing crank. This crank arrangement is now universally used on six cylinder in line four stroke engines. On later engines, incidentally, R increased the main bearing size to 1.75″.

A moral to be learned from this episode is that if you copy you must be certain that what you are copying from is satisfactory. If not, then correcting the result will take longer and cost more than making sure in the first place. An interesting sidelight is that A.J.Rowledge, who crops up later at Rolls-Royce and then became known as Rg, was in April 1903 designing the Napier L48 six cylinder engine, 6.25″ by 5″.

A page of crankshaft stress calculations made by him, dated the 10th of that month, shows quite clearly the now standard six cylinder crank arrangement. A car fitted with this engine achieved a speed of 104.65mph on 24th January 1905. Rg had more time to look into both principles and details than had R in those hectic three or four years.

Electric vehicles were enjoying a brief popularity for town use. R was asked in the spring of 1905 to design a town carriage as an electric vehicle "look-alike" (68, 69). To add to the difficulties, the engine had to be vibrationless, noiseless, require little or no gearchanging, be smokeless and free of unpleasant smells.

Because it would be helpful to put the engine as far away from the passengers as possible, R decided to put it under the driver's feet rather than under his seat, as others had done. The resulting requirement for a very shallow engine was met by a V-8 configuration with 3.25″ bore and stroke and side valves. This combination gave an overall height only about 5″ more than the flywheel diameter. The force path from cylinder head to crankshaft bearings was shortened and made more direct by putting the main bearings in the upper half crankcase; and the need to eliminate oil smoke meant changing from a total loss to a circulating oil system. This allowed oil to be provided under pressure to the main and big end bearings. Connecting rods were kept simple by side by side big ends. Overall length was kept down by a three bearing crankshaft, of the single plane configuration (70, 71, 72).

The car layout had the advantage of providing a dash to back axle dimension of 103.5″ on a wheelbase of only 90″. By comparison, the 'long' 30hp on its wheelbase of 118″ had a dash to back axle figure of 84″. An associated car of more normal layout, the 'Legalimit', so called because it was governed not to be able to exceed 20mph, had a wheelbase of 106″ and only carried a two-seater body. Only three cars were made with this engine, one of them a 'Legalimit' intended for Lord Northcliffe but not retained by him. The Landaulet had a frame width of 37.5″, the 'Legalimit' 33.5″. The latter was the same as the 'Light 20', the T.T. car. These three exceptions to the 36″ frame width shared the 52″ track of the 15hp model.

The concept was dropped but useful lessons had been learned. A redesign of the 30hp engine, with larger main bearings in the upper half crankcase and pressure fed bearings was never actually made. The ability to mount ever larger bodies was becoming a requirement. The resulting overall weight of such vehicles needed more power than the 30hp engine could provide.

A new start was therefore decided on. Although Montague Grahame-White, who had a considerable influence on bodywork styles at the time, wanted a wheelbase of 150″, the eventual decision was for alternative wheelbases of 135.5″ and 143.5″. More significantly these gave dash to rear axle figures of 85″ and 93″. Although both were more than the 84″ of the long wheelbase 30hp they did show the effect of a longer engine and a more normal engine position behind the front axle.

The new engine, the 40/50, still had seven main bearings, now in the upper half crankcase and with oil pressure fed to main, big end and little end bearings. There were two measures to increase the torsional stiffness of the crankshaft: main and big ends were increased to 2.25″ from the final 1.75″ of the 30hp, and a "square" engine had 4.5″ bore and stroke. The shorter stroke reduced the equivalent rotating inertia of the reciprocating masses and so further raised the torsional frequency, putting the more troublesome vibrations outside the normal running range. There were two cylinder blocks each of three cylinders, incorporating the cylinder heads, so that there was now no restriction on the length of the centre main bearing, the most heavily loaded. All valves were now alongside the cylinders and operated from a single camshaft by rocking levers, broadly as in the V-8 engine. The lower half crankcase and oil sump were combined. Magneto and coil ignition were provided and the Royce carburetter further improved, and its governor control linkage, were incorporated. The gear drive to the camshaft and auxiliaries and the new centrifugal water pump, with impeller and casing cast in bronze, were at the front of the engine (73, 74).

There was less difference in the chassis. Platform rear springing continued; the external contracting footbrake still worked on the transmission, and the handbrake on the rear axle drums, using for the first time I believe a small bevel gear differential to equalise brake efforts side for side. All brakes had two shoes per drum and worked equally well in both directions. Engine and four speed gearbox were separate and the engine mounting was arranged to isolate it from chassis distortion. A complete car (75) and a polished exhibition chassis were on the stand for the 1906 Olympia Show. R felt the strain of these considerable achievements and had to take odd days off.

A 40/50 christened the "Silver Ghost" by CJ, after a first class performance in the Scottish Reliability Trials of June 1907 went on to do a 15000 mile endurance test (the maximum acceptable) under R.A.C. observation (76). A petrol tap which vibrated shut caused the only involuntary stop; it took 30 seconds to find and rectify. The remaining 14,331 miles were trouble free, a record at the time. After the trial only £2.13 had to be spent replacing parts not still within the manufacturing tolerances. Tyres, at 1.25p per mile, cost twice as much as petrol, oil and all repairs. Rolls-Royce Motor Cars Limited still own this car. The name is now generally applied to the model.

Rolls-Royce Limited had been formally registered on 15th March 1906; a slightly reluctant (because he did not want to leave Manchester) EAC was elected as its first chairman.

Royce Ltd and the directors of Rolls-Royce Ltd held the small part originally issued of the authorised capital of £60,000. The first public issue financed early in 1907 the complete absorption of C.S.Rolls & Co. but only the car making part of Royce Ltd at Cooke Street. The Rolls-Royce Ltd capital was soon increased to £200,000 to finance plans for a larger factory. The Trafford Park works of Royce Ltd continued to make cranes; R was seen fairly often at the London sales office until at least the very late twenties.

Over nearly three years Cooke Street had turned out about three cars a month, had done all the development work and the laborious preparation of the four T.T. cars. A forecast of between six and seven cars a month for the first ten months of 1907 was too few for the demand immediately generated by the 40/50. It was sensible to limit production to this model; a further advantage was that this would also limit the strain on R.

The Board wanted the ability to produce 200 cars a year. Early in January 1907 funds were voted to acquire machinery for this, and then CJ and R were told to find and buy a site for a new factory. They chose one at Nightingale Road, Derby. A factor in the choice was the availability of a reasonably priced electricity supply. R got the job of laying out these works (77, 78, 79, 80, 81, 82), specifying the new equipment needed and organising its installation and the move from Cooke Street without loss of production. Not surprisingly he had to take time off as a result but was back in time for the official opening on 9th July 1908.

He had rented "The Knoll" at Quarndon, north of Derby, and moved there with his wife Minnie and Violet Punt, her niece and later their adopted daughter (12); they also had a lodge, "Seacote", at Overstrand near Cromer where they spent weekends.

With the 20hp no longer available for racing, CSR (45), transferred his enthusiasm to ballooning and flying. He was tragically killed in an accident at a flying display at Bournemouth in 1910, a month after making the first double crossing of the English Channel on his Wright biplane.

With R's efforts channeled into a single model there were continual developments. In 1908 the platform rear springs were replaced by 3/4 elliptics, effectively only a rearrangement of the spring material. These were followed by cantilever rear springs in 1911, associated with a change to a torque tube rear axle, connected to the chassis by a patented spherical joint. On this axle too there were separate and concentric brake drums for foot and hand brakes; the foot brake used the outer drums (83).

In late 1909 R's continuing parsimony showed itself in a reversion to his pet three speed gearbox. With a 'crash' gearbox many drivers disliked gearchanging and for them this change might be regarded as an advantage. The associated reduction in the ratio range between top and bottom gears however caused a 40/50 entered by James Radley in the 1912 Austrian Alpine Trials to stop on the Katschberg pass. It restarted when relieved of two of its passengers and completed the climb. Three official cars in the 1913 event, with four speed gearboxes, swept the board. Radley again scored a victory in 1914.

An exception to the Rolls-Royce 'no racing' policy, which was adopted when the 40/50 became the only model, occurred in 1913. Spain announced a Grand Prix race for fully-equipped touring cars (84). The results would be based on a handicap, including petrol consumption; bonnets were to be sealed and cooling water could not be added during the race. Count Carlos de Salamanca (85), who became the company's Spanish agent, thought the 40/50 could win this and persuaded the company to prepare two cars. They came first (driven by Salamanca) and third (driven by Platford). The race was over three laps of a twisty and mountainous 64 mile circuit which rose to a maximum height of 2,500 ft. Of the 17 starters only 11 finished. By that time the chassis had cantilever rear springs, torque tube rear axle, double kingpost frame stiffeners and the now 4.75″ stroke engine (83).

R's health got worse. He was operated on in 1911 at Norwich after falling ill at Overstrand, and nurse Ethel Aubin was hired, on a temporary basis, to look after him. After a period of convalescence at Overstrand they went in August by train to France, met CJ and were taken to embark on a cruise to Italy and Egypt, returning in the spring of 1912 to the south of France. R had originally been given only a few months to live, but had recovered well. To maintain this improvement he was advised to remain in a warmer climate and follow a more relaxed routine; the combination of the two would mean that he would have to operate away from the works at Derby. This was a relief to many because his uncompromising attitude on quality had become an embarrassment, not least because he was liable to sack people if he felt they were not meeting his own standards of workmanship. At that time the French Riviera was popular because of the mildness of its winter climate. According to Wilton J. Oldham, 'The Hyphen in Rolls-Royce - The Story of Claude Johnson', CJ already had a hillside villa at Le Canadel (86, 87), and R was so enchanted with the view from it down to the sea that he decided he could happily work there. CJ seized the opportunity and acquired land to establish quite a colony. Below CJ's Villa Jaune there were to be R's Villa Mimosa, Le Bureau for the design office and Le Rossignol to house the designers (88, 89, 90). Whatever the original intention, the final one was that R should spend the summer in the south of England, within more reasonable reach of the works, and the winter at Le Canadel. In pursuance of this CJ brought R back to a house, 'Westwood', at Crowborough while preparations were continuing at Le Canadel.

The situation at 'Westwood' was unhappy. There was friction between nurse Aubin, now permanent, and Minnie and Violet. Minnie had a horror of illness, but Violet, not so handicapped, felt excluded by the limited access permitted by nurse Aubin. Patient and nurse departed in October to Le Canadel for the winter, initially at the hotel because the house was not yet finished. R's troubles, however, were not yet over. A recurrence of cancer of the bowel led to his being rushed to London by car for a colostomy operation.

When he was sufficiently recovered R went to convalesce at 'Westwood'. With nurse Aubin now more firmly in control the situation was worse. R and nurse left soon for Le Canadel, the final separation between R and wife Minnie and adopted daughter occurred. Minnie and Violet stayed on at Westwood until 1921. R ensured that Minnie was provided for until she died; he also, later, gave Violet a handsome wedding present. Nurse Aubin's reign lasted until R's death in 1933.

5. **Aero Engines, World War 1 and Onwards**

The Le Canadel colony established in 1912 was so comprehensive that it must have been intended as a permanent home. R and his staff however for the summer of 1914 were installed in a house, 'Seaton', at St. Margaret's Bay, between Dover and Deal and not too far from another of CJ's villas. This may have been because it had been realised that the south of France was too remote for reasonable communication with the works at Derby and the to-ing and fro-ing of the various people who had to see R, and the cars he wanted to see and try, or it may then have been a purely temporary precaution to make sure R was safe and accessible if war did break out. The exact reason is immaterial; he was there for the summer. The war, when it came, would make it advisable for R to stay in the south of England. They remained therefore at 'Seaton' until cross-channel shelling increased enough to make their position dangerous. Late in 1917 R and Miss Aubin went househunting westwards along the Sussex coast. The last house they looked at was Elm Tree Farm (91) in West Wittering; Maurice Hewlett had moved there earlier from Wiltshire when lack of manpower suspended his farming activities there. He was now anxious to move back to Wiltshire because labour was becoming available from men invalided out of the services, but still capable of doing farming work. R liked the area, noted for its mild winters, and moved there as soon as possible. It was established during the war that R could certainly stay in the south of England during the summer, and it would be a sensible precaution for him to spend the winter and early spring in the south of France once the war was over and that became possible again. A routine was therefore established on this basis.

It was at 'Seaton' therefore that James Radley visited R to receive his congratulations on his success again in the Austrian Alpine Trials. Radley was a reserve officer in the RNAS section operating airships, and during the visit he and R were watching one of these airships struggling against a headwind and achieving very little progress because of the very inadequate power of its engine. This led to Radley begging R to let him have a 40/50 engine, suitably modified. He was convinced that this would do better than their existing equipment. This was the second time that R had been asked to agree to the use of a car engine for aircraft use. CSR had made the earlier suggestion when he bought his Wright biplane and thought the 20hp four would be better than either the Wright engine or its European substitutes. In both cases R turned down the idea. To us now the idea of using such heavy engines, with cast iron cylinders, may seem absurd, but even the Sunbeam V 12 of about that date had cast iron cylinders and side valves. R might have turned down the idea of the 20hp because he thought the lubrication was not good enough; he had progressed to force feed to the crankshaft bearings by that time on other models.

His attitude changed when the war started. So did the Rolls-Royce Board's panic decision to close down the works completely. The 40/50 was quickly modified to carry the weight of an armoured car body; many normal ones became staff cars. Rolls-Royce were also asked to make the Renault V-8 and RAF 1a engines, both air-cooled. R criticised the design of both, especially the former. If they were to make aero engines they might as well make something they approved of. R decided to design one himself. When it was ready they would switch production to it.

A period of intense activity followed. From Commander Wilfrid Briggs of the RNAS, who was later to recruit W.O.Bentley too, R learned of requirements for 100, 150 and 200hp engines and decided to start on the last. He assembled the relevant factors. Thermal problems with the cast iron pistons then in use made him limit the cylinder bore to that of the 40/50, 4.5″, and insist on water cooling, rejecting the strong suggestion of air cooling from Mervyn O'Gorman of the Royal Aircraft Factory. A lighter construction was necessary; examples of this, and of overhead valves to improve the breathing, were available in the Mercedes aero and car racing engines at the Derby works, so he made his only visit there after his 1911 illness, to study them. They provided the idea of individual steel cylinders with welded on steel water jackets. R decided on two valves per cylinder in a part spherical head, operated by a single overhead camshaft and rockers. The 40/50 engine gave about 80hp. To get the power he would need two banks of six cylinders, in Vee form, a stroke lengthened to 6.5″ from the 4.75″ of the 40/50, and a speed of about 1800 rpm. To keep the propeller blade tip speed down to an acceptable figure, for the size needed for this power, a reduction gear would be necessary.

By 11th September 1914 arrangement drawings of the main engine (probably 0.25 scale) and a six cylinder derivative to meet the 100hp requirement were available. A lot of detailed scheming remained to be done, including the epicyclic reduction gear, chosen to avoid increasing the loads on the crankshaft front bearing. Aluminium pistons, recommended to E.W.Hives (Hs) by W.O.Bentley, on the basis of his pre-war car racing results, were eventually adopted. This feature had been the original reason for W.O.'s recruitment by Commander Briggs.

This design is an excellent example of how to achieve a target by limited steps from a known main basis of proved reliability. The result, known as the Eagle (92, 93, 94) was first tested by Hs (95) in February 1915. Its 20.32 litres soon gave 225hp at 1800 rpm. This was increased to 360hp by the war end.

Conflicting requirements from the Navy and the War Office, including suggestions of no reduction gear from one, and a projected requirement for detachable exhaust valve seats for maintenance, which R was convinced would not work because of the inferior heat path, had to be sorted out and this did take time. The dimensions of 4.5″ by 6.5″ were arrived at by 8th September. R wanted the reduction gear for the twelve cylinder engines to save weight overall by the combination of higher engine speeds and smaller cylinders. He proposed no reduction gear for the six cylinder version because of fears of torsional vibration.

There was a quite early suggestion of smaller cylinder dimensions to meet approximately the 150hp requirement. The final decision on that was for 4″ by 5.75″, to be achieved by scaling down from the larger engine. Most of this work could be carried out at Derby under R.W.Harvey-Bailey (By). The resulting engine was called the Falcon of 14.21 litres (96, 97). Its 205hp in April 1916 became 228 in May and 285hp at 2000 rpm by the end of the war. A single bank of this, but with 6″ stroke and no reduction gear, made the 7.41 litre Hawk for airship propulsion (98, 99). This gave 91hp in February 1916 and finally 100 to 105hp at 1500 rpm. In spite of apparent complication - 50% more parts than comparable engines - all these engines were noted for their reliability.

The 35 litre Condor (100) appeared in 1918, had no war service, and was in construction a scaled up Eagle but with four radial valves in each part-spherical cylinder head. Series 1A and 111 each gave 650hp at 1900 rpm but the latter was 300 lb lighter. Part of this saving was due to the use of a spur reduction gear in place of the epicyclic. This also gave a better disposition of the frontal area about the propeller axis.

R insisted that they should stay in the aero engine business after the war. His enthusiasm for the Condor up to 1924 was not shared by the Air Ministry, which favoured the single block per bank of the Birkigt-designed Hispano-Suiza and the

Curtiss D-12 and wanted Rolls-Royce to be the maker. After preliminary scheming under R the detail designing was handed over to a design office at Derby under A.J.Rowledge (Rg), who, under Montague Napier, had been responsible for the Lion before moving to Rolls-Royce. By June 1927 this engine, the F10, had passed its 100hr type test, naturally aspirated, at 490hp and 2350 rpm at the first attempt (101, 102). Supercharged versions followed. It was named the Kestrel in 1930. Final developments of the Kestrel were the Goshawk, a steam cooled version, and the Peregrine (103) which on 87 octane fuel had a maximum output of 885hp at 3000 rpm and 15,000 ft. It had some Merlin features and a downdraught carburetter. The Vulture (104) effectively used four Kestrel cylinder blocks on a common crankcase and with a single crankshaft. It had a two-speed supercharger and at 3000 rpm had maximum powers of 1845hp at 5000 ft, low gear, and 1710 at 15,000 ft, high gear. The last two engines saw limited service.

The Buzzard or H-engine, 6″ bore and 6.6″ stroke, 36.7 litre, was a 1.2 to 1 scale-up from the Kestrel (105). It had a normal rating of 825hp at 2000 rpm, a maximum of 955hp at 2300 rpm, both at sea level. This was the basis of the R-engine, victorious in the 1929 and 1931 Schneider Trophy races (106). With a bigger supercharger it gave 1900hp at 2900 rpm for the 1929 race. For the 1931 race a target of 2300hp at 3200 rpm was set, to be held for one hour on test. From 20 minutes in April it moved by stages to the target on 12th August at 2350hp. A 'Sprint' version in September gave 2530hp at a specific weight of 0.64 lb/hp. In recognition of his part in the earlier success R was made a baronet in June 1930, becoming Sir Henry Royce of Seaton in Rutlandshire.

The Merlin design, known then as the PV12, was started in January 1933 and the first engine ran in October of that year (107, 108). It had twelve cylinders 5.4″ by 6″ and a capacity of 27.02 litres. In 1937 the Merlin II was rated at 950/990hp at 2600 rpm and 12,250 ft altitude, with 890hp at take-off at 2850 rpm. World War II saw too many developments to list here. It was the last aero engine started in R's lifetime and was the mainstay of the RAF effort in the last war, in fighters and bombers. It is dealt with in 'The Merlin in Perspective', number 2 in this historical series.

6. Cars Between the Wars

Immediate moves after the Armistice was signed on 11th November 1918 were to tail off aero engine production and to get the peace time version of the 40/50 going again. A smaller model to complement this had been decided on, in view of the much increased works capacity and a realistic expectation of little more than development orders for aero engines; the RAF would be living on its fat for years. A first sketch of the bodywork and layout of the new car was finished in four days and a further three weeks produced more detailed proposals. A general arrangement scheme dated 13th March 1919 implies that West Wittering had finished the bulk of the design work.

The aero engines had shown the advantages of overhead valves and the first engine, codenamed 'Goshawk', used two overhead camshafts to operate them (109). The drive to these was too noisy so a single camshaft in the crankcase replaced them, with push rod and rocker operation of the valves (110, 111, 112). Size reduction was found to result in little cost saving. A redesign to save more had R's pet three speed gearbox, in unit with the engine and with centre change; rear axle torque was taken by semi-elliptic springs (113, 114, 115, 116).

There would be other savings from a smaller car. It would be suitable for an owner-driver (116); and the annual tax of £1 for each RAC rating horsepower (Rating HP = $D^2N/2.5$, with D the bore in inches and N the number of cylinders) would be £22 a year compared with £49 for the 40/50; and there would be savings on petrol, oil, tyres, maintenance and insurance. The car was introduced in 1922 and immediately sold well. It had 3″ bore and 4.5″ stroke and a capacity of 3.13 litres.

Rolls-Royce had first entered the U.S. market in 1906. CSR took over the winning T.T. car and two 30hp sixes. The Light 20 T.T. car won a race for 25hp cars just before the December Annual New York Motor Show and was sold to Captain Hutton who entered it for the Daytona and Ormond Beach races and at the latter set up a world 5 mile record for cars up to 60hp. At the Show orders were taken for the six and an agent appointed. Then on a visit in 1911 CJ met two Canadian cousins of R, Allan Henry, a lawyer, and James Charles, a qualified engineer, both of whom served the company in their respective fields (1). James Charles Royce was appointed "Engineer for the USA and Canada" later that year. CJ was considering the possibility of manufacture there, perhaps with two factories as close as possible, one each side of the water dividing the two countries. This idea was followed later by several American car companies. No immediate action was taken, but Rolls-Royce did buy a factory at Springfield in 1917, thinking then of aero engine manufacture. CJ again went over in June 1919 and Rolls-Royce of America Inc. was formed in November to make Rolls-Royce cars there, mainly to avoid the very heavy import duties then levied on cars. Initial difficulties, and a suspicion that Springfield cars weren't as good as the Derby originals, were overcome. 1700 of the Ghost cars were made, compared with 6173 built in England, mostly of course at Derby, and over a much longer period.

Production at Springfield was also eventually changed to the New Phantom which, as mentioned later, had succeeded the 40/50 Silver Ghost at Derby in 1925. Many of the bodies for both models were made by Brewsters, which had been bought by Rolls-Royce for this purpose.

Back home, an example of the continual model improvement was the announcement in late 1923 that front wheel brakes were to be fitted to the 40/50; actual fitting began in 1924. Argyll had incorporated them on production cars before 1914 and they were practically universal on racing cars from that time. Operating twice as many brakes meant twice as much effort and in 1919 Hispano-Suiza fitted a mechanical 'servo' to help the driver. The 1922 TT Vauxhall had compressed air to apply the brakes. R's first effort used it only to help rod-operated brakes; Hs found it to have uncontrollable vibrations. Although West Wittering continued to work on mechanical details of the brakes, including an alternative to the Perrot system, which R disliked, for the front (117), it was left to Rg and Hs to sort out the mechanical servo (118). By worked on an idea to avoid the Hispano-Suiza patent but this was not adopted. Although Rolls-Royce paid licence fees to Hispano-Suiza their design owed nothing to the patentees.

In May 1925 an overhead valve engine, fitted to the existing 'Ghost' chassis, was announced and called the 'New Phantom'. Its 4.25″ bore gave it an annual tax of £44, its 5.5″ stroke put up the capacity from the final 7.43 litres of the Ghost to 7.67 litres (119). Its extra developed 20hp was useful to deal with the ever heavier bodies which were being fitted (120). Unlike the new 20hp car it still had two blocks of three cylinders; the common cylinder head stiffened the structure; a single plate clutch was fitted.

For the initial chassis weight of 18 cwt for the 20hp R had specified an all-up weight not exceeding 35 cwt (116); this was not held. From the beginning half the owners wanted enclosed bodies with divisions (121). The car was attractive for town work; its suitability for this was increased from 1925 by the addition of a four speed gearbox, to give the ratio spread needed for the extra weight, and of four wheel brakes for safety in traffic. For many years Rolls-Royce headed the London traffic census figures.

1929 saw two more changes; the 3″ bore of the 20hp went up to 3.25″ to make it the 20/25 (RAC rating 25.4), better able to cope with still increasing weights (122, 123); and there was a new chassis for the 40/50, which now became the Phantom II (124, 125). Rear axle torque was no longer taken by a torque tube but by the semi-elliptic springs which replaced the cantilevers used on the final version of the Silver Ghost and the New Phantom, which thenceforward was referred to as the Phantom I. Engine and gearbox were no longer separate and the drive from the combined unit was by way of an open propeller shaft to a hypoid axle, the Gleason company's improved version of spiral bevels. In this case there was a 1″ downward offset of the pinion shaft from the crownwheel axis.

A crossflow cylinder head on the engine (126) with a six-branch partially water-heated inlet manifold on the right or driver's side and a three-branch exhaust manifold on the left, with a lagged downpipe, gave more power and less noise. The engine was rigidly mounted, probably to increase frame front end stiffness; the frame side members were stiffened by the then common double kingpost and tie arrangement and had a slight drop in level between axles to give a lower body.

Front axle brake torque was taken by a parallel link arrangement using front-shackled springs and damper arms (127), the latter connected to the axle by triangle links whose bases had fore and aft pivots to allow roll and single wheel bump. Single diameter rear brake drums had hand and foot brake shoes side by side. Centralised chassis lubrication fed all the frame parts; there was a single feed point on the front axle for its needs.

This new chassis put paid to Springfield, though Rolls-Royce of America Inc. hung on to 1931. The 1929 slump in America ruled out any hope of recovering the tooling cost. Maurice Olley (Oy), who had been chief engineer there, moved to Cadillac late in 1930 to undertake a programme of suspension development which led to the general adoption of independent front suspension ('knee action') on U.S. cars for the 1934 model year.

The ever-increasing weights of the 20/25 cars led to two experimental engines, codenamed in typical fashion Japan II and Japan III, intended to provide more power. The first had a longer stroke of 5″ (compared with 4.5″), in conjunction with the 3.25″ bore of the 20/25. The second had a larger bore of 3.4″ and the same 5″ stroke. The second certainly, and the first probably, had a feature new to Rolls-Royce car engines, a combined crankcase and cylinder block. There was also a new chassis associated with them. Neither went into production.

A further scaling-down exercise, from presumably JII, produced the Peregrine. I was told at West Wittering that the scaling down ratio was 0.85. That ratio from 3.25″ gives 2.7625″, and Peregrine had a bore of 2.725″, intended to keep the RAC rating below 18hp; 2.7625 would have given 18.3. The stroke also was reduced from the direct ratio, to 4.125″ rather than 4.25″, to give a capacity of 2.36 litres. It was said to develop over 70bhp on a compression ratio of 6.0 to 1. Six were built. One of the engines, like its parent with a combined crankcase and block, in cast iron, can be seen at Hunt House, the home of the Sir Henry Royce Memorial Foundation. It was a lively and likeable car, but because it was again found that the scaling down gave little saving in cost, and because it would probably have taken some sales from the 20/25, it was dropped.

A sporting derivative with a lower chassis was however the basis of the Derby Bentleys. The 8 litre Bentley, thought by many to be a better car than the Phantom II (even the Continental version, a partial answer to it) was the only car the original Bentley company had in production when the 1929 slump arrived. The resulting drop in sales led to the appointment of a receiver in 1931 and by some last minute manoeuvring Rolls-Royce acquired the company, together with W.O.Bentley himself, legally one of the assets of the company. The name was used on a 'Silent Sports Car'. Supercharging the Peregrine engine, the first idea, was dropped because of gasket trouble and a cross-flow head 20/25 engine then under development replaced it. The car was announced in 1933 as the 3.5 litre (128). Compared with the 20/25 this engine gave more power, the chassis was 2 cwt lighter and the 6″ shorter wheelbase (at 10′6″) limited bodywork to four seaters. It gave an excellent performance (129).

The front axle tramp on some Phantom IIs and 3.5 litre Bentleys prevented soft front springs being fitted. Therefore either the whole suspension was too stiff for comfort or a tendency to pitch existed. Oy had a theory, no doubt known to R and now generally accepted, that the rear suspension had to have a higher periodicity than the front to avoid pitching. Independent front suspension, which permitted softer front springs and therefore provided a better ride, was tried experimentally on a Phantom II (27 Ex) to designs started at West Wittering in 1932 incorporating equal and parallel transverse triangle levers (130). Independent front suspension, with about double the spring track of a front axle, would have about four times the anti-roll stiffness; in this case this was reduced to about two and a half times as much by the softer front springs. Even so, the torque applied to the front end of the frame by single wheel bumps showed that it was now much too weak torsionally. Also, the bump and rebound movements originally provided were not enough, partly because of the softer springing and also because with independently sprung wheels the bump movement for single and two wheel bumps is the same, whereas with a front axle the narrow spring track allows single wheel bumps much more than the parallel bump limit. Finally, the effect of a two-piece cross steering tube with each tube considerably longer than the triangle levers introduced steering effects with up and down movements of the wheels which produced unacceptable 'joggles' at the steering wheel. Nevertheless, the ride was so much better than that of a car with a front axle that some form of independent front suspension was felt to be necessary on the new model, codenamed 'Spectre', whose design was started in 1932 and which was finally introduced in 1935 as the Phantom III. The massive frame and its cross were boxed throughout (131). The length from frame cross to front wheels was minimised by: a twelve cylinder Vee engine potentially 6″ shorter (3.25″ bore instead of 4.25″); a separate gearbox behind the frame cross intersection; and swinging the now unequal length triangle levers forward at their inner ends. The long lower links gave good steering geometry with a two-piece cross steering tube; short upper ones left room between them for the front of the engine and the radiator (132, 133, 134).

An experimental car (Vulture) with Phantom II engine bored out to 4 3/8″ and 25hp (Japan III) gearbox and hypoid back axle showed that the last two components were adequate and confirmed that the A.E.Berriman argument of 1929 for the Daimler Double Sixes, that large six cylinder engines had individual punches too big for refinement, was correct.

Present day owners grumble about the complication and servicing problems of the Phantom III. Albert G.Elliott (E), aero and car chief designer, and Bernard I.Day (Da), in charge of chassis design under him, had worked for R for many years, so that his influence, even if diluted in parts, persisted not only in the Phantom III but also in the Wraith, the 1938 replacement for the 25/30 (the bore was 3.5″ by now), in which some attention was paid to cost and serviceability. The frame, boxed and provided with a boxed cross, was lighter and easier to make and stopped short at about the front wheel centre line (135). The front suspension was a unit; a tail from each spring housing was held by four bolts to the frame member inside which it fitted (136, 137, 138). Engine and gearbox were in one; there was a hypoid axle. The front suspension geometry avoided the excessive gyroscopic effects of the early Phantom IIIs, which was due to last minute changes from steel to wire wheels and altered sharing of suspension movements between bump and rebound. These geometry changes were embodied in the D series Phantom IIIs. The complete Wraith brake linkage (139) shows that the basic principles of the original, 1924, layout incorporating the servo had been maintained, but with changes to suit chassis developments, such as independent front suspension, and the somewhat relaxed legal requirements for the handbrake.

The Wraith wheelbase was 6″ less than that of the Phantom III but the front wheels to dash dimension was less. The tracks were mid-way between those of the 25/30 and the Phantom III. The weight was 9 to 10 cwt less than that of the larger car. The Wraith therefore had very nearly the accommodation of the Phantom III but less performance. Allowing for the weight and gearing it should develop some 82% of the tractive effort per ton of the Phantom III; the maximum speed would inevitably be less. It can be seen as a first step towards the "Rationalised Range" which was the basis of the Rolls-Royce and Bentley post-World War II cars.

7. **Coping With R's Separation From The Works**

From the time of his first bowel operation in 1911 until he died in 1933 R was separated from the works. The primary reason for this separation was the need for a warmer climate and a more relaxed routine; the secondary reason was to avoid the spur-of-the-moment dismissals R was liable to indulge in if he thought some poor workman was failing to keep to his own high standards. Even if there was not quite the immediate pressure that a fast-moving production line implies, cars had to be produced in reasonable numbers to support the organisation. Unexpected, and possibly unjustified, interruptions were a nuisance and could be expensive.

For most of that period of separation R remained in active control of the design of the product and the product engineering set-up had to be arranged to get the most out of R without overstraining him. The choice of designers to work with R was important. More will be said about that later. To keep the strain on R to a minimum he would preferably work with as few designers as possible. Although they would have to be able to work on their own after an initial discussion, if R's attention turned then to other matters, his designers would normally expect a visit by R at least once a day, when they would show and explain what they had done. They must then be willing to make whatever changes were required. An india rubber was essential and local thinnesses of the tracing paper produced by it were a hallmark of the successful design scheme. The required combination of ability, patience and persistence was not easy to find.

Getting the full benefit from his designers meant that they had to produce design schemes only. This often tended to produce the typical crowded design scheme, with odd views, notes, possibly tables and calculations all squeezed in, so that everything would be clear for whoever had to produce detail drawings (140).

When it was ready the original design scheme, on transparent tracing paper, would be sent to Derby with an issue note on which any further information could be included and which would specify what the scheme was intended to do and what action should be taken once the detail drawings were finished and their associated list of parts (LOP) prepared. The original scheme would be retained at Derby but a blue print would be sent back so that it could be worked from. One of the nuisances of the wet process blue prints then in use was that they shrunk, so that if you wanted to trace from one, as for instance to provide the scenery for another job close to it, you would have to shift the print under your tracing in order to maintain correct relationships at widely separated areas. In due course too blue prints of the detail drawings would be sent to West Wittering and incorporated in the filing system there.

These detail drawings, one for each individual part, fully dimensioned and with working tolerances on those dimensions where necessary, were made in the product drawing office. This, with its associated jig and tool drawing office, had to remain at Derby. Tolerances on critical dimensions are chosen in the first place by functional requirements and they then determine the methods of manufacture. Slight relaxations may sometimes make manufacture easier; the possibility of these needs discussion between the product drawing office and those who plan the works processes. There are also inevitably queries from the works about the parts and these also have to be answered by the product drawing office. When I knew it, this office was very capably headed by R.W.Harvey-Bailey (By - 22nd February 1876 to September 1962) (141). After schooling at Tiffins in Kingston-on-Thames he won an exhibition to King's College, London, finishing up with an AKC in civil engineering in 1896. He was however attracted to motor engineering, working as a draughtsman for Col. Pennington at the Horseless Carriage Co. Having left there to join Coulthards at Preston on steam engine design, he did some design for Dr. A W Brightmore, Head of the Engineering Department of the Royal Indian Engineering College, on a steam driven front wheel drive phaeton. The power unit was concentrated in a two wheel bogie, incorporating a differential: steering was by braking the inner wheel on a corner, to deflect the bogie in the right direction. This apparently required a good deal of skill and By proved to be very good at it. He persuaded Dr. Brightmore that the arrangement was more suitable for a heavy steam wagon, and it was arranged that By should carry out the designs for this, still at Coulthards. Its novelty was a drawback. By then moved to the Critchley Norris Co., still in Lancashire. Norris had been at Coulthards too, and an early job for By was converting the Brightmore wagons to rear wheel drive so that they could be sold. Another job was the design of an omnibus using a 30/40hp Crossley engine. There were associations not only with Crossley but also with the Peter Pilkington Co. which made steam hammers. By had ideas for these also but resigned when Critchley Norris laid claim to them.

He joined the Pullcar Co., at Preston in Lancashire, which made front drive petrol engined cars, and was there involved in the design, contemporarily but coincidentally with Christie in America, of a transverse (White & Poppe four cylinder) engined front wheel drive car with an epicyclic three speed and reverse gearbox. This was intended for a taxi, which became the first one to pass the new police regulations in 1908 (142). These for the first time incorporated the still current 25′ turning circle requirement. The driver sat right at the front, with the engine and gearbox under his seat; the larger rear wheels were directly behind the door opening and there was a 3/4 landaulet opening top. The driving unit was also applied to a delivery vehicle aimed at the Post Office. Pullcar did not obtain either order. By wrote to Rolls-Royce for a job but had no reply, and took a job with the Pilgrim's Way Motor Co. Here again, and in this case against his wishes and recommendations, he was required to design a front wheel drive car with an air-cooled flat twin engine. He managed to achieved a single pedal control for the three speed and reverse epicyclic gearbox, one pedal less than the model T Ford which had only two speeds and reverse. Although this worked well it was not marketed. By eventually joined Rolls-Royce in May 1910, after a year at the Pilgrim's Way Co. The initiative for this came from CJ. He was put onto a weight reduction exercise on the new 40/50 Silver Ghost torque tube axle, a noise reduction campaign for the crown wheel and pinion (instigated by Arthur Wormald (Wor)) and went out with R to Le Canadel in October 1912, completing his design for the four speed gearbox required for the 1913 Austrian Alpine Trials and doing some site clearance on the side. He was to remain with Rolls-Royce until his retirement at the age of 69, at the end of 1945. He was 85 when he died; he was certainly still driving at the age of 80. He presented a paper, "Front Driving of Steam and Petrol Engines", to the Institution of Automobile Engineers (Proc I.A.E. Vol II, 1907-08).

In the organisation I knew By had an assistant, Bill Clough (Cgh) and four section leaders heading Engine, Frame, Gear and Electrical sections, producing detail drawings with E, F, G and D prefixes respectively.

By had applied himself to the question of the materials to be used for the parts. These were listed and their characteristics were tabulated. There were also established practices to be followed where possible. There were lists also of standard parts, bolts, nuts, washers, spring washers, etc., and catalogues of small parts, levers for instance, to be looked through to see if an existing one could be used rather than create a new one.

By's department also did design work when required; most of the design of the Falcon aero engine was done there, effectively scaled down from the Eagle. This function however was largely but not wholly taken over by A.J.Rowledge (Rg) (143), who joined Rolls-Royce in 1921 from Napier, and his designers. Rg greatly improved the Condor and, after initial contacts with R and his designers, the work on the F10 (later known as the Kestrel) was also his responsibility. Visits to R were frequent, so the design evolved in full consultation with him. Because of the time scale involved, most of the design work on the 'R' engine adaptation of the Buzzard for the two Schneider Trophy races in 1929 and 1931 was also Rg's responsibility, after full discussion of failures in development and agreement on probable cures (106). It should be noted, however, that the H engine or Buzzard, was designed under By's control as a scale-up of the Kestrel, and that R instructed the design with potential for Schneider Trophy applications in mind.

The design scheme might be a modification of an existing model or a part of a completely new one. In both cases the result would have to be tried for behaviour, and if that was acceptable for endurance too. Endurance testing of cars was eventually done in France, at a small establishment set up by W.A.(Roy) Robotham (Rm) (144) in 1924 at Chateauroux. The site was chosen as providing a good variety of driving conditions within daily reach. Operating a two shift system allowed endurance mileage to be completed far more quickly than in England, and good relations were established with the local constabulary.

Originally, the regular endurance test on each car was 10,000 miles; it was eventually, and a good deal later, increased to 15,000 miles. Cars to be tested might have various new features, or be completely new. In the first case one 10,000 mile test might be regarded as a sufficient check on the novel features. For new models three such tests would probably be necessary. Specification changes, or new models, could only be introduced on the basis of the results of these tests and any such changes had to be agreed by a collection of department heads including designing, detailing, experimental, works and sales departments and the managing director himself. Dire penalties, were threatened to anyone who bypassed this procedure.

While he was at the works R was not only in charge of design: he was also chief development engineer and kept a watchful eye on tool designing and manufacturing equipment. From 1911 the role of development or test engineer fell naturally onto Ernest W.Hives (Hs - 1886 to 1965) (145), though he was not officially made head of the experimental department until 1914. He was a cycle mechanic in Reading when he was discovered by CSR, for whom he worked personally and also in the CSR service department. He left there, before the fateful R - CSR meeting in May 1904, to go to Napier. He moved on from there also, among other things did some driving at Brooklands, and was eventually taken on as a tester by Rolls-Royce at Derby late in 1908. Soon after this R had taken him on a test drive to the Cat & Fiddle; on the way back to the works R took him home to Quarndon, and asked him if he was hungry. He replied: "Sir, I could eat a bald baby with measles!". R described him as a "coming young man". He organised the London-Edinburgh run in 1911, managed the magnificent and victorious official three car team in the 1913 Austrian Alpine Trials. Testing the Eagle (95), Falcon and Hawk aero engines was taken in his stride. When R's first effort at servo-assisted four wheel brakes produced shattering judders, it was the combination of Rg and Hs which in 1924 produced acceptable equipment. He moved up to be General Works Manager in 1936 when Arthur Wormald (Wor) died and was appointed to the main Board in 1937.

Hs had collected several able assistants. A.Cyril Lovesey (Lov) was an early one, who eventually specialised on aero work, as did James E.Ellor (Lr), the supercharger expert. W.A.Robotham (Rm) (144), after four years as a premium apprentice, was appointed as Hs's technical assistant in 1923 and took on most of the car work. An early effort was the setting up of the car endurance testing organisation in 1924 at Chateauroux, chosen for having all types of going within reach of the two shifts a day testing. He was a forceful and energetic character and when car and aero engineering were separated in 1937, with E as chief engineer, aero, and By as chief engineer, car, Rm was appointed as By's assistant. Running ahead well beyond R's lifetime, as an indication of the characters of those who had come under R's influence, By, Bernard Day (Da) and Bill Hardy (Hdy) were abstracted from the car division at the start of the war in 1939 to help in aero engine work. A nucleus of the car division was left under Rm's charge, with Ivan Evernden (Ev) as chief designer, initially to tidy up existing car work, and eventually, thanks to Rm's recognition of the dreadfully chaotic situation of British armoured fighting vehicle equipment, to become design and development parents of the Cromwell cruiser tank, which brought about completely new standards of mobility, with its Meteor engine (a naturally aspirated Merlin), reliability and fighting power. Hs became the mainstay of the tremendous Rolls-Royce effort in World War II. He was made a Companion of Honour in 1943, Lord Hives of Duffield in 1950. He retired as chairman of the Rolls-Royce Board in 1957, and died in 1965. His story is dealt with in "Hives : The Quiet Tiger" in the Memorial Foundation and Heritage Trust series.

8. **R's Designers**

Any typical engineering manufacturing company has a drawing office. The abler draughtsmen would be given design jobs and would then expect, with the help of juniors as required, to be responsible for the preparation of the resulting detail drawings, probably doing the major components themselves. R was familiar with this type of organisation and used it. As we have seen, the separation of R from the works resulted in, and was made possible by, the selection of a specialised band of designers too limited in numbers to be employed in this way. Although By was an early example of such designers, as we have seen, it was soon realised that his overall qualities and experience could be better utilised at Derby in the way already described. R's early team therefore consisted of two designers with complementary abilities, Albert G.Elliott, (E), and Maurice Olley (Oy) (146). E relied heavily on his practised eye, Oy much more on his abilities to analyse and calculate. It is surprising how quickly such a combination can work. For instance, between 4th August (the declaration of war) and 11th September 1914 they effectively finished the Eagle design layout and made arrangement drawings of both the main engine and a possible six cylinder derivative. This was after R had found out the requirement and decided how to meet it.

E was born on 3rd December 1890; between 1905 and 1908 he studied, on a sandwich course, at the Northampton Institute in Clerkenwell, and was awarded a diploma in mechanical and electrical engineering. From 1908 to 1912 he worked at D.Napier and Son, Acton, mostly on car engine design. He joined Rolls-Royce in 1912, very soon started to work for R and remained with him, at Le Canadel, St. Margaret's Bay and West Wittering (147), until December 1931, when he moved to Derby, some 16 months before R died. He was therefore established at Derby before the rest of R's designers were transferred there after R's death. His colouring rather suggested the fox and I thought he had some of the other characteristics too. He was R's favourite until he moved to Derby. He had been appointed chief designer in 1929, chief engineer, aero division, in 1937, to the Board in 1951, later became joint managing director, and executive vice-chairman in 1954. He was a member of the Institution of Automobile Engineers and served on its Council from 1936 until 1947 when it became the Automobile Division of the Institution of Mechanical Engineers. He was also a fellow of the Royal Aeronautical Society. He was made a C.B.E. in 1941, for his work on aero engines; and was awarded the British Gold Medal for Aeronautics in 1954 by the Royal Aeronautical Society for "outstanding practical achievement leading to advancement in aeronautics". He retired from Rolls-Royce in November 1955, though he remained as technical director of Rotol Ltd. until September 1963. While at West Wittering he lived at Wickersfield House (147), at Derby he lived at 'Holmwood', Quarndon; he later lived at Stratford-on-Avon, more convenient for the Rotol premises. He died on 13th November 1965, just failing to reach his 85th birthday. Rotol Ltd. had been founded jointly by Rolls-Royce and the Bristol Aeroplane Co. to design and make variable pitch propellers.

Oy was born on 12th June 1889 in Scarborough. He got his matriculation from the Victoria University of Manchester in 1905 and from 1906 to 1907 did one year of the Honours engineering course there, unluckily ended by bronchitis and pneumonia. This led to a year on a farm, living and sleeping in the open. From 1908 to 1911 he was apprenticed to H.W.Ward in Birmingham; during 1911 and 1912 he was a tool designer there, doing special jobs. He joined Rolls-Royce in November 1912, working for Thomas Nadin (Na) on plant layout. He moved to Le Canadel to work for R in April 1913, initially on tool and jig design on milling fixtures for machining connecting rods all over, then on the design of a starter and generator. Though back at Derby for a brief spell, he returned to Le Canadel in October 1913 and stayed with R from then on. They were at St. Margaret's Bay from July 1914 onwards, apart from a spell at Bognor from December 1914 to April 1915, during which E was at Derby. In July 1917 he was released to go to the USA to discuss Eagle and Falcon manufacture there. With short intermissions he remained there, doing various jobs for Rolls-Royce, culminating in being chief engineer to Rolls-Royce of America Inc. at Springfield until October 1930. On 1st November 1930 he joined Cadillac to continue work he had started at Springfield on ride improvements. American road conditions showed up the fundamental pitching tendencies of the Rolls-Royce Ghost and New Phantom. Oy realised the need for rear suspension frequencies to be higher than the front. To make this possible independent front suspension was necessary and his work led to the general adoption of "knee action" in America for the 1934 model year. He remained with General Motors until he retired in June 1957 at the age of 68. He spent most of the thirties at Vauxhall in England, exercising a considerable influence on suspension improvements and other developments. He returned to the United States at the beginning of the 1939-45 war and was on loan to Rolls-Royce during that period. He maintained close contacts with friends at Rolls-Royce while at General Motors but refused an invitation from Hs in 1937 to become chief engineer of the car division there.

The oldest of the West Witteringites was Bernard Incledon Day (Da) (1879 - 1956), (148). He was the son of John Sage Day, initially of Dublin. He was educated at Hartlebury and Presteign grammar schools, followed by Birmingham Municipal Technical School and Sheffield University. He married Evelyn Russel, the daughter of Dr. John Foster. She was as kind-hearted as Da and a doer of good works. From 1908 to 1913 Da was Chief Designer of the Sheffield-Simplex Motor Works and responsible there for the 45hp six cylinder car of 4.25″ bore and 5″ stroke, which was said to have caused Rolls-Royce some anxiety because of its high quality. He also lectured in automobile engineering at Sheffield University from 1910-13. He then joined Rolls-Royce and during the 1914-1918 war was in charge of its Airship Design Office and was there responsible for the installation of the machinery in most of the early British rigid airships. After the war ended he headed an office at Derby doing chassis design work, under R's direction, and when it became possible in 1921 he moved to West Wittering and remained there until the move to Derby in May 1933. He was responsible for patents on such varied subjects as friction clutches, swivelling propellers, to vary the direction of thrust by as much as 180°, and aircraft landing gear. He was never one to push himself forward but was accepted as the chief chassis designer. R had a great respect for and trust in him. He was a first class designer of high grade mechanisms, and the kindest and most considerate of men. For me, he was a pleasure to work for and an ideal mentor.

He was elected a member of the Institution of Automobile Engineers in 1916 and was a Fellow of the Royal Society of Arts. He and his wife lived at Ingle Nook, West Wittering, before he built his bungalow, 'Cloverfield', (149). It was typical of him that he had built a bungalow rather than a house so as not to intrude more than was necessary into the seaward view of any of the houses in the village. He and his wife gave frequent tennis parties on their grass court during the summer. When I arrived at West Wittering he had a Windsor car; I think this was a 10hp by then aging but with the right shape of radiator for the quality car it was. It was soon replaced by a more mundane Austin 12-6.

William Guy (Bill) Hardy (Hdy: 26-8-1891 to Dec 1979), (150, 151, 152) first worked at Clement-Talbot at Barlby Road in London and was elected a graduate of the Institution of Automobile Engineers in May 1913. He spent the war, initially in the RNAS, then in the RAF when this absorbed the RNAS and the RFC, in what became the Aircraft Inspection Directorate. After a period near Birmingham he was moved to Derby, carrying out his duties at Rolls-Royce there. He returned to Talbot when he was demobbed. He had got to know several of the Rolls-Royce people during his work at Derby and managed to get taken on there in 1919, working under Da in the drawing office on chassis design work. He was transferred from graduate to associate member of the I.A.E. in October 1919. Da was moved to West Wittering in 1921 when the Camacha studio became vacant. At that time Elmstead was fully occupied by E, Charles L.Jenner (Jnr) (153), C.E.Stent and Charles Carolin. Stent too had been at Derby in the RNAS. He was chiefly remembered in the village for his painstaking restoration of a Humberette car. Hdy also moved to West Wittering in 1921, for a trial period of 3 weeks. After three years he was told he could build a house there, and the thatched 'Clematis', in keeping with its neighbours, was the result (154).

The Camacha designers (H.Ivan F.Evernden, Ev, became the third) worked only on chassis design. Hdy became the transmission specialist. He thought the Wraith gearbox the best of the pre-World War II synchro-mesh gearboxes, better than the Phantom III (140). E and Jnr, the surviving occupants of the Elmstead design Office, did car engine design work and were also involved in project aero engine design, which was then taken over by A.J.Rowledge (Rg) at Derby.

Hdy commented later on the fact that the West Wittering designers, when they moved to Derby in 1933, were respected rather than resented; and on the shift of power from design to experimental departments after R died, noticeable before the last war and even more marked after it. Hdy was a thorough and thoughtful designer and very much a perfectionist. He became chief designer of the car division after the war, until he retired, then going back to the West Wittering area. Both he and his wife Jane seemed in good health at the unveiling of the 'Camacha' plaque in September 1979, so that it was a shock when they both died, within a few days of each other, in December 1979 (151).

Hdy still had a Redwing Riley when I started at West Wittering but this was fairly soon changed to a 1.5 litre six cylinder Riley saloon. That in turn was changed for a four cylinder 1.5 litre Riley because of the poor petrol consumption of the six. He continued to live at Duffield, just north of Derby, until he retired, commuting to Crewe after the car division moved there.

H.Ivan F.Evernden (Ev: 11-8-1896 to 1980) (155), was called up in 1916 before finishing his engineering degree course at King's College, London. He failed his medical and was directed to Rolls-Royce at Derby, working in the drawing office. He too was moved to the Camacha studio in West Wittering in April 1921 and eventually became responsible for styling and liaison with the coachbuilders. He was one of R's favourites. During World War II he was in charge of the car division design office at Clan Foundry in Belper. As mentioned earlier, the car division nucleus became the design and development parent of the Cromwell, a fast cruiser tank of some 29 tons embodying Christie type suspension, the Merritt-Brown transmission incorporating a much sounder method of steering track-laying vehicles than the then normal clutch and brake, and the unsupercharged Merlin engine christened the Meteor. Ev was awarded the M.B.E. in 1943 for his work in this connection. When I arrived at West Wittering in 1932 he had a not very reliable 14hp Delage tourer; this was soon changed for a second-hand 20hp Rolls-Royce tourer (156), for which he had acquired chassis and body separately. He arranged for one of his coachbuilders, I believe Barkers, to combine the two. The result may not have been as dashing as the Delage but was much more reliable, more prestigious, and very appropriate to a Rolls-Royce designer.

Jnr, mentioned earlier, had started his working life at the Pilgrim's Way Motor Co., working under By, in 1908 or 1909. It was By who offered him a job at Rolls-Royce and later arranged for him to work for R at West Wittering, where he effectively became

E's assistant. His name first appears in the LeC design register there in February 1918. He was taken to Derby by E in December 1931. Although he had originally worked on aero engines he eventually specialised on car engines and was finally responsible for all car engine design, including the overhead inlet, side exhaust engine, conceived and made experimentally before the last war as a part of the rationalised range, but finally developed after it to become the commercially successful B range of engines, with four, six and straight-8 cylinder versions. Although Jnr had the overall responsibility for this much of the major scheming was done by Alec Mitchell, Jnr's junior, recruited at Derby. His part in the Cromwell tank episode was the adaptation of the Merlin to derive the Meteor. A confirmed bachelor, he was another very kind and considerate character and a first class designer steeped in his craft.

Although Rg (30-7-1876 to 12-12-1957) (143), never worked at West Wittering I feel he should be mentioned in this collection of notes about R's designers. He was born and grew up in Peterborough (R's period at the locomotive works began when Rg was a year old); like R he was the youngest of his family of three boys and one girl and his mother came of Lincolnshire farming stock. He developed wide interests, art and science particularly, and was apprenticed in his home town to Barford & Perkins, engineers, until 1898. Short experiences in printing machinery and heavy engineering in the London area led to a post as designer with D.Napier & Son in 1901, on cars. I have a copy of a page of stress calculations he did on the Napier 'Samson' racing engine, six cylinders 6.25″ by 5″, some 15 litres capacity, dated April 1903, which is interesting as showing the now accepted crank arrangement. He left to join the Wolseley Co. in Birmingham in 1905. In June of that year he married Alice Blincko, of Kennington; they had three children, a girl and two boys. During a good part of the time at Wolseley he was chief designer. In 1913 he returned to Napier as chief draughtsman and was later chief designer. He was involved in the design of high class cars there. Like Rolls-Royce, Napier originally in World-War I made aero engines designed elsewhere, by the Royal Aircraft Factory and Sunbeam. Rg learned enough from these to start in 1916 the design of the broad arrow, three banks of four cylinders, water cooled Napier "Lion" engine, based upon a study of the requirements of fighter planes; it incorporated a spur reduction gear so that the propeller axis was roughly at the centre of area. By 1918 this engine, of 5.5″ bore and 5 1/8 stroke, 31.925 litres capacity, in Series II form was developing 450hp.

Rg was responsible for the design of the T75 post-war Napier car, which did not sell very well. When the design was blamed for the poor sales Rg decided to leave, and in 1921, at the age of 45, he joined Rolls-Royce as R's chief assistant, with the title of Chief Aero Design Engineer. He was stationed at Derby, but did make frequent visits to West Wittering for consultations and discussions with R. An early task was to increase the Condor power from 540 to 650hp and reduce the weight drastically. Part of this exercise was the incorporation of the single spur reduction gear, similar to that of the 'Lion'. As we have already seen, having come to the conclusion that the potential for future development of the Condor was limited by the inherent flexibility of the structure with separate cylinders, he took on the design of the F10 or Kestrel engine, after some preliminary discussions with R, which had a single six cylinder block per bank. He played a significant part in the further development of the Kestrel's larger brother, the Buzzard, as the 'R' engine for the 1929 and 1931 Schneider Trophy races. In part for this work he was awarded the Gold Medal of the Institution of Automobile Engineers in 1931, and the first award of its Simms Gold Medal in 1932. In 1936 he was awarded the British Silver Medal of the Royal Aeronautical Society for "scientific achievements in the development of aircraft engines". After a serious illness in 1933 he was appointed Chief Consultant to the company, and continued to be involved in a variety of developments until his retirement in 1945 at the age of 69. He was a quiet and thoughtful man, applying basic principles and a study of requirements and problems to aircraft engine design. He had a very considerable influence on the company's technical progress.

Others who worked at West Wittering were Samuel F.Pottinger (SFP), later to work for Rg at Derby, and Don Eyre (DE), a talented artist who gave painting lessons to R (89, 90) and could be relied upon to produce lightning sketches whenever the fancy took him. When designing a two speed supercharger, for instance, he told me that it was "for climbing steep clouds" and illustrated this idea on the edge of the cartridge paper cover of his board. I wondered later if there was any significance in the ability of three of R's favourites, E, Ev and DE to produce 'good looking' schemes. As I have mentioned, E fell from grace when he moved to Derby in December 1931, taking Jnr with him; he had acquired a previous black mark for marrying R's secretary, Miss Gray. For a separate establishment such as R had at West Wittering good secretarial assistance is essential, with a great deal of a Girl or Man Friday in the make-up. Miss Gray's replacement was 'Monty' Marmont (148, 152) who typed memos, sent them off, ran errands and generally did all the odds and ends assosicated with a detached group; there were the farming interests as well. He had a bicycle for dashing between Elmstead and Camacha and other local journeys, including initially getting to and from work. He was both good natured and efficient, and cheerful too. He had a liking for the name 'Locksash' for his home (157). The second house bearing this name was built on land sold to him by R in Rookwood Road, nearly opposite the end of Elms Lane, and therefore much closer to Elmstead.

The fact that these very senior designers continued to work on the drawing board was unusual but was necessary to fit in with R's special requirements. I believe however that the first essential of the Rolls-Royce technical success was excellence of the original design. The second was the refining of the process of development springing from R's passion for improvement. Using the minimum amount of material for any mechanism had always to be kept in mind.

W.O. Bentley once remarked to me that a particular designer was too much of a pessimist to be good at his job. A designer certainly must believe that by the time he has finished it his design must work as he intended it to. The complete engineer must however accept that weaknesses do occur and must be prepared to do something about them. R was an excellent example of this.

Hs thought that R's designers had become so used to being told what to do by him that after his death they made more mistakes than decisions. I feel that this opinion can be understood from someone with his essentially practical early training, but in making it he fails to realise that engineering is essentially the art of compromise. Whatever we may want to do is beset with disadvantages of various kinds to weigh against the advantages we are looking for. The closer we can get to the best compromise at the design stage, the quicker and cheaper will the final development period be. This is even truer today when computers can both speed up and make more accurate the theoretical assessments, but it was certainly true before such aids became available. As soon as one gets to the actual embodiments both costs and time spent escalate.

An example of ill-judged experimental work was mentioned to me by Jnr. One of the American cars being evaluated in the experimental department had the first example seen at Derby of the tuned harmonic crankshaft torsional damper, which has a rubber spring in the form of a ring between the hub and the damping flywheel. It was very effective on the engine to which it was originally fitted. The damper when fitted to a Rolls-Royce engine was not effective. The conclusion was drawn from this that this type of damper was not good enough for our engines and that we should take no further interest in it. Whoever initiated this experiment did not know that the harmonic damper has to be tuned to the critical frequency of the engine to which it is to be fitted. Unless this is done it is impossible to compare its effectiveness with the then more general friction damper which works in a different way. After E had taken Jnr to Derby with him in December 1931, at the same time as R went off for his usual Le Canadel stay, it was decided that the three remaining West Wittering designers, Da, Hdy and Ev, should each take a junior. I was lucky enough to be appointed as Da's junior. A.E. (Tony) Cooke (AEC) was Ev's and we were all four at the Camacha studio (158). Harry Biraben (151) (HEB), was Hdy's junior and they worked at Elmstead. On R's first visit to Camacha, with his black Labrador 'Rajah', I got a strong impression of shyness or reserve. There were tales of how he would use, or fabricate, almost any excuse to avoid, at the last moment, going to functions even, or perhaps particularly, when he was going to be an honoured guest. As a last resort his health could be used to explain his absence. This sounds very different from the impression of an overbearing, intolerant and demanding individual that Bill Morton* gives for the R of later Cooke Street and early Derby days, even though lightened by the touches of humour and the instances of kindness to others that are quoted. Age could account for some of the changes; a major difference was that in the early days of the car venture R was having to re-establish himself and the company in its new field and would not spare himself or others until its position was secure.

R's past achievements in the electrical field would have justified even arrogance, and the knowledge of how much still needed to be done would account for impatience and intolerance of anything less than the maximum ability and effort on the part of himself and those near him. When I knew him his and the company's positions were firmly established, there was a team of men on whom R knew he could rely, and it was up to him, and Nurse Aubin (159), to decide on how much effort he could put in. His work remained his one compelling interest.

An interesting indication of his attitude to the use of non-working time, even then, was that he met Tony Cooke carrying a bathing costume. After the usual polite greeting he said: "I hope you're not going to do anything frivolous, Mr. Cooke". There might have been an implication that swimming for health-giving reasons might be acceptable, like R's walks with Rajah, but purely for pleasure less so.

Another indication of R's attitude to non-working time was given by Hdy. When he was intending to go away for the week-end R would be very liable to engage him in conversation, mostly on trivial matters, at the end of the Saturday morning period, and after keeping him for a long time would dismiss him, saying: "Off you go, if you must, but don't drive too fast and kill yourself".

Hdy also commented on the shift of power after R's death from design to development, and more particularly after the war. He thought this tended to lead to slower and more expensive development of new designs. In his opinion the West Wittering designers, after they had arrived in Derby, were respected rather than resented, though apparently not by Hs!

---

* 'A History of Rolls-Royce Motor Cars, Volume I, 1903-1907' by C.W.Morton, published by G.T. Foulis & Co.

9. **Other Colleagues**

Because of the numbers involved it would be impossible to refer here to any but the most significant of these. R provided the technical foundation of F.H.Royce and Co., Royce Ltd. and Rolls-Royce Ltd. and a fair amount of commercial common sense. He depended very considerably upon others for the commercial organisation which is so essential to overall success. He needed, and had, people with complementary abilities to his own. The first example of these is certainly EAC (13), the 'Lord High Everything Else' of F.H.Royce & Co. It seems likely that it was because he recognised R's technical ability that he was willing to gamble his £50. The choice of name for the company supports that idea. It would then be reasonable for EAC to take the precaution of being on the spot to see that everything possible was done to make the most of R's ability. He provided physical support and encouragement of R in the very early days, and took on all the other activities needed to prop and build up the firm.

EAC was the son of a London doctor; he was born in 1864, had had some electrical training at what became the Brush Electrical Co. and had two other jobs before joining R at Cooke Street in Manchester. He had a fetish for physical fitness and was something of a martinet. He was fond of tennis and gymnastics in pursuit of this fitness. He expected an evening meal when he arrived home after some such exercise; his wife, Minnie Royce's sister, was clearly very house-proud and by that time was more interested in putting dustsheets on the furniture for the night, and there were clashes. As mentioned elsewhere, the marriage did not last long. EAC found it easier to command a housekeeper than to argue with a wife. He had managed the F.H.Royce & Co. finances very well and was respected for this by the local business community, to the extent that he was invited by Henry Edmunds to join the W.T.Glover board. This was on a reciprocal basis, with Henry Edmunds joining the Royce Ltd. board. EAC became the first chairman of Rolls-Royce Ltd. in 1906 and remained in that position until he died in 1921.

Two others may be mentioned as early recruits to F.H.Royce & Co. Ltd.: John De Looze (D), accountant and company secretary, continued as such through the various company and title changes until well after R's death. Arthur Wormald (Wor) was a trade union activist before being taken on; presumably appointment as shop foreman changed that. He was eventually General Works Manager at Derby until he died.

There are some notes about CSR in Chapter II. CJ was brought in by CSR; his life is thoroughly covered elsewhere*; a few facts about him are appropriate here. He was born on 24th October 1864 at Datchet in Bucks, the sixth in a family of seven. His younger brother Basil (BJ) succeeded him as managing director of Rolls-Royce but was in no way his equal and was soon replaced. St. Paul's School and the Royal College of Art - briefly - took him at nineteen to the Imperial Institute as a clerk. This was through the good offices of Sir Philip Cunliffe-Owen, under whom he worked in organising a series of exhibitions with increasing responsibility. He was made chief clerk to the committee arranging for the building of the Imperial Institute at the instigation of the Prince of Wales as a tribute to the Queen on her Golden Jubilee in 1887.

The Institute was officially opened on 10th May 1893. CJ was given by the Prince of Wales the job of organising an 'International Exhibition of Motors and Their Appliances', in association with the Motor Car Club. This was held from May to October 1896, to allow the cars to be demonstrated as well as exhibited. It was here

---

* 'The Hyphen in Rolls-Royce', by Wilton J. Oldham, G.T.Foulis & Co. Ltd. now a part of the Haynes Publishing Corp.

that the Prince had his first car ride. The successor to the Motor Car Club was the Automobile Club of Great Britain, formed in 1897; CJ was appointed its first Secretary in December of that year. It was through this that CJ came to know Alfred Harmsworth, later Lord Northcliffe (85) and they became life-long friends. Although the first trials organised by the ACGBI (Ireland had been added in 1898) in June 1899 were financially disastrous, CJ proposed a much more ambitious project for the following year, the '1000 Miles Trial'. The Committee was understandably cautious about this, but Harmsworth, who had founded the 'Daily Mail' in 1896, supported the idea, promised it wide publicity and agreed to finance it. It was a great success. CSR won a Gold Medal in it, driving a 12hp Panhard. The Club prospered, CJ was very active in promoting acceptance of motor cars and looking after the interests of members. He was overworked, and left in 1903.

CJ, by way of a fairly short stay with Paris Singer's City & Suburban electric car project, accepted late in 1903 an offer from CSR to join him as a partner in C.S.Rolls & Co. which had been founded the year before. The two men were an ideal combination. CJ was the obvious choice for effectively the managing director, though not originally so in name when Rolls-Royce was founded in March 1906; he remained so until he died in April 1926. His influence was profound. He steered the company towards the one-model policy, partly because of his concern for R's health, but also because it was commercially sound. After R's operations, in 1911 and 1912, he was concerned that the company should have a possible successor to R. He caused By to be engaged. He looked at the possibility of James Charles Royce, R's engineer Canadian cousin, and later made a much appreciated but refused approach to Birkigt of Hispano-Suiza. It is possible that Da's move from Sheffield-Simplex was also a part of this policy. Rg's recruitment from Napier could also be seen in this light, though the initial approach could well have been Rg's. CJ certainly realised that thoroughly competent engineers were essential to the company's continuing development and prosperity.

EAC may have known, or more probably known of, Henry Edmunds (37), the company engineer at what became the Brush Electrical Co. while EAC was training there. The partners, R and EAC, will certainly have known him for some time as a result of dealings with the W.T.Glover electrical cable firm. The knowledge and mutual respect of Edmunds and EAC led to each becoming a director of each other's company, and therefore of Edmunds' increasing knowledge of R and his abilities. Others have indulged in imagery by calling Edmunds the godfather (one of two) of Rolls-Royce because he brought about the meeting of R and CSR. By doing so he played an important part in the creation of that firm and thereby also introduced CJ (85) who has been referred to as the hyphen in Rolls-Royce. I think the original idea of making cars was R's, recognising the weaknesses of all the cars he had come in contact with, sure of his own ability to devise something better than these cars, and aware of the need for other projects to keep the existing Royce Ltd. factories fully occupied. So R would have proposed making his original three cars anyhow. If the meeting with CSR had not taken place R would certainly have stayed with his original 10hp twin much longer. My suggestion of imagery therefore is to suggest that Henry Edmunds' function was that of the midwife. This leads naturally to the idea that CSR then supplied the fertility drug which caused such a rapid increase in the family. Think how rapidly the twin was followed by the three, four, six, eight and finally the Super Six. Following this idea still further, CJ becomes the obstetric consultant who realised when the mother (R), exhausted by so much childbirth, should be advised to confine her future activities to the nursemaids and governess' roles!

Developing this idea still further, Harry Briggs could be the true godfather, prompted by the faithful retainer D to round off the subscription list which provided the funds for the more commodious nursery premises to allow the last, and only surviving, child to come to her full development!

An alternative scenario might represent R as Cinderella, Henry Edmunds as the briefly occurring Fairy Godmother and CSR as Prince Charming. In that case CJ would become the Prime Minister, Conduit Street would become the Cabinet Office and the general Rolls-Royce staff the members of parliament. EAC, the company chairman, would be the Speaker of the House.

Certainly the Board of directors and the London sales and financial organisations need to be mentioned collectively for the important part they played in the continuing existence and prosperity of the company.

In this general context of the way in which apparently trivial incidents can have far-reaching effects on the lives of others, it is interesting to think that but for CSR's extraction of Hs from his cycle shop at Reading he would probably not have been available in 1908 to join Rolls-Royce and play such an important part in its later development.

10. **Life at West Wittering**

The sixteen years that R spent at Elmstead in West Wittering provided his longest spell in any one home; granted that he spent about four months of each year in his Villa Mimosa at Le Canadel, but Elmstead can be regarded as his more settled home during that period. His immediate staff would certainly regard it as home. Over the greater part of that time, from 1914 to 1930, the Rev. Arthur Tidman Gill was the vicar of West Wittering. Among the eleven surviving children of his wife Rose and himself (160) were Eric Gill the sculptor and Macdonald (Max) Gill the architect and artist, always associated in my mind with his delightful map of Kensington Gardens, used as a poster by London Transport. It was Eric Gill who carved on R's mantelpiece the motto: "Quidvis recte factum, quamvis humile, praeclarum" - the Latin translation of the motto "Whatever is rightly done, however humble, is noble" which R had justifiably either adopted or had conferred upon him. Eric Gill no longer lived in the village when I was there, but Max Gill had a house, near to the old coastguard cottages, where he lived with his family, and worked. One of his daughters still lives in the village.

West Wittering is a small village about eight miles from Chichester and at the western end of the Selsey peninsula. It had no gas or electricity. It had in 1931 a population of 695, roughly 200 more than it had had thirty years before. Most of them were I think still true villagers, though some of the cottages had been taken over as weekend or holiday homes: Evelyn Laye, for instance, had one of these. There were also several retired people, some others, and at least one author permanently living there. Rolls-Royce were responsible, in 1932, for six designers, one secretary, and four wives as well as R himself, Nurse Aubin and whatever domestic staff were employed at Elmstead - say some fourteen or fifteen people.

A map of the village and its surroundings is helpful (157). The old harbour near the coastguard cottages, mostly silted up, was enclosed by a ridge of reed-grown sand dunes. From the Point, at the end of this ridge, you looked across a narrow gap to Hayling Island. There was quite a tidal race through this gap, the entrance to Chichester Harbour. A good deal further away, to the south, you could on a clear day see

Bembridge in the Isle of Wight. Flashes from the Nab Lighthouse could be seen at night.

Within a few yards of each other could be found the pub, 'The Old House at Home', Adcock's Garage, and Trickey's store, which jointly provided the focal point of the village. There were new houses here and there; Da and Hdy had both built. There was also a row just behind the shore, along The Strand. While we were there one of these, with a thatched roof, got completely burnt out, leaving only walls, chimney stacks and a roof water tank propped up on its pipes. The fire engine had to come from Chichester, taking some 15 to 20 minutes, and was too late to help. There has been a good deal of building since 1933, as became obvious at the unveiling of the commemorative plaque on the Camacha studio wall in September 1979 (161, 162). One noticeable change in the meantime is the building of the New Road to the car park behind the beach. Even so, West Wittering remains comparatively undeveloped. East Wittering was a more popular resort, Bracklesham then somewhat more exclusive.

The studio in the garden of Miss Ramsey's cottage, Camacha, had either been converted from an old barn or specifically built as a studio. From 1921 it had accommodated Da, Hdy and Ev. In 1932 it housed Da and Ev, Tony Cooke and myself. Hdy had moved to Elmstead and Harry Biraben was there too; so was Monty Marmont. At Camacha we had to rely upon an earth closet, and with no electric light we each had an incandescent mantle oil lamp for working after dark. This was hung above the top left hand side of the drawing board. The cylindrical wick needed frequent attention. Hdy and Harry Biraben were luckier than us in both respects; they had electric light from the house supply of 24 volts from a generating set and batteries. In addition to the four drawing boards and the tables to take the drawings to which we needed to refer, there were a lot of plan filing cabinets for blue prints of our scheme drawings and the detail drawings of the current and some past models, with their associated bound LOPs, on blue print linen or paper. I think the boards were 30″ by 40″, to take with some margin the normal maximum drawing size of 26″ by 38″. They had moving straight edges across the board, controlled to run parallel by thin multi-strand wires at each end. These ran on helically grooved pulleys at the top, firmly fixed to the ends of a transverse spindle, and on single grooved idlers at the bottom. A sample can be seen at Hunt House, the home of the Sir Henry Royce Memorial Foundation. We covered the boards with cartridge paper to provide a white backing for the transparent paper on which we worked, cut off as required from a roll. Standard issues for each designer were an adjustable set square (most of us had fixed set squares providing 30° + 60°, and 45° - provided by ourselves), a 12″ nickel silver Chesterton rule, pencils of assorted hardnesses, an india rubber and a thick coconut fibre mat to stand on - so that if you dropped a pencil you probably wouldn't break the point and so would not need to waste some pencil and your time re-sharpening it. We also had a red crayon pencil for scheme titles: numbers top and bottom on the right, scale (mostly 1/1 or 1/4) at the bottom left and title across the bottom. We had a duster for flicking off the pencil dust from the drawing, and we could have narrow sandpaper pads for getting exactly the right shape on the pencil lead - point or chisel edge. Most of us preferred a file for this, hung by a string from the bottom left hand wire pulley bracket.

R's designers had two scheme drawing registers: one recorded drawings with an LeC prefix, whose origin is obvious; the other, drawings with an N Sch prefix. I don't know the basis of that. E took the LeC register with him to Derby in December 1931. Derby also had a scheme drawing register, with DES prefixes (Derby Engine Schemes). I believe this was started when A.J.Rowledge joined the company.

In the summer sea bathing and tennis were fairly obvious ways of spending some leisure time. Tony Cooke and I were lucky because Atherstone House, opposite Elmstead in Elms Lane, where we had 'digs', had a hard tennis court. Da had a grass one at 'Cloverfield'; he and Mrs Day gave evening or week-end tennis parties. For bathing there was a first class sandy beach stretching all the way to Selsey Bill. Tony Cooke and I bathed at least once a day in the summer, sometimes twice - he was by way of being a fresh air fiend. These dips went on well into October, and the first the following year was early in April - and very cold. There were plenty of local walks; a favourite one was to the Point.

In the winter there was badminton in the village hall (designed by Max Gill) two evenings a week. Several oil lamps were hung round the sides as high up as possible. The rather complicated timber roof trusses added interest to the game. High clearing shots could usually find their way through; if not a let was played. Some of us went over on Sunday afternoons to play in a hall at Bognor. We played occasionally as club guests in Chichester. We sometimes went to a cinema there too; I remember going to a performance of Gilbert & Sullivan's 'The Yeomen of the Guard' by the local operatic society.

Both Tony Cooke and I were recruited from being post-graduate pupils at Daimler in Coventry. The first indication to both of us that we might be coming to work at West Wittering was when I saw his CMS Morris Minor Special outside the studio on the Saturday morning when I attended for an interview. I had an Austin 7 'Chummy', later changed for a Fiat 8 (156). We were lucky both to have cars, because they made life in West Wittering much more bearable for young bachelors, when used for journeys into Chichester and week-ends away. There was a bus service into Chichester, but it was not particularly frequent or convenient.

R's death in April 1933 meant the end of our life at West Wittering. We gathered on the lawn at Elmstead to witness the departure of the coffin for cremation. The absence of ceremony and fuss would, I am sure, have pleased R; the only stranger I can remember being present was Mr. Tyldesley from Claremont Haynes the solicitors, though I am sure there must have been others.

It took a little while to organise our own departure; a place had to be found and prepared for us in the design office at Derby. We finally left on 23rd May at the end of a day of packing and what farewells had not already been taken (152, 156).

11. R Himself

This section contains some speculation as well as fact and hearsay. Up to the age of fourteen R endured both poverty and hardship; during at least the latter part of this period the family was held together and mostly supported by his mother, Mary (King) Royce, with such help from her children as they could provide. Towards the end R spent some holidays with the aunt at Fletton, as a result of which he had his just-over-three-year spell as an apprentice at the Great Northern works at Peterborough. R must have expressed some interest in engineering, or shown some aptitude; possibly also his mother remembered that both his father and his grandfather had installed steam power in water mills. The aunt was probably his mother's sister. From 1877 to 1880 therefore R was better fed than ever before. More important in many ways was that practically all his time was being spent on educating himself and working towards being able to earn a reasonable living in an occupation which interested him and for which he had an aptitude.

Apart from providing him with pocket money, R's newspaper round at Peterborough may have been to pay for his English and mathematics evening classes. His single-minded devotion to his job first appeared during these three years. Luckily his spare time efforts with Mr. Yarrow bore fruit when his apprenticeship came to its untimely end. A normal period would have been at least five, possibly seven, years. In spite of being only seventeen, he was taken on as a reasonably qualified toolmaker by Greenwood & Batley. There in spite of a fifty-four hour working week, and some overtime - he was often required to work through Friday night - he managed to learn enough about the new-fangled electricity to get his next job. I don't know whether he had to move because the Italian Arsenal work was over or whether R heard of a job going with the London Electric Light and Power Co. and went for, and got it. He knew his mechanical ability by that time; he needed to use the theoretical knowlege he had been busily gaining, and considerably add to it. Where better could he do this than in London? It paid off. As first electrician to the Lancashire subsidiary in Liverpool he had his first taste of being his own technical master, and discovered that responsibility brought anxiety too.

The one thing he had had no time for, up to the age of twenty one when he was again without a job, through no fault of his own, was fun. It was EAC, after they had set up in business together, who provided a little of this, either during or after such meals as they could afford. Otherwise it was still work, as long as he was able, or awake.

It is over the next nine years that the greatest change occurred in R, from a young mechanic and electrical engineer, embarking on a risky business partnership, to an assured and mature professional engineer in full control of the technical side of a thriving business. He was on the verge of marrying Minnie Grace Punt. Because his partner EAC was also about to marry Minnie's sister, and between them the two sisters were putting £1500 into F.H.Royce & Co., I strongly suspect that these were arranged marriages, probably very soon after the death of the father of the two young ladies concerned. Mr. Punt's firm had undertaken printing jobs for the Royce business, eventually for Rolls-Royce.

Shortly before this R had brought his mother to live with him at 45 Barton Street, Moss Side. Throughout his life R showed gratitude and loyalty to those who had helped him. In the case of his mother there was a mutual affection too. This was some time between 1889 and 1893; his mother had been at first housekeeper, finally nurse to a gentleman with a substantial establishment at Tunbridge Wells. Her employer died, she was about sixty. It was about time for R to return some of what he owed her. As the youngest of her children, either he cared more than his older sisters and brother, or was in a better position to look after her - or perhaps both. When he married and went to live in Chorlton-cum-Hardy he installed her in lodgings nearby and used to visit her on his way home from work. The marriage was in 1893; realising the wealth of experience R had by then had, and the size of the business in relation to the minuscule beginning nine years before, it was something of a surprise to me to realise that R was only 30 when he married.

An arranged marriage can develop happily. A catalyst, helping to put some warmth into it was provided by Minnie's niece Violet Punt. Minnie had a brother living near Johannesburg. When his first wife died the two children suffered; Violet at least ran wild. She was a high-spirited, open air tomboy, able to ride any horse, usually bare-footed, feeling free by wearing as few clothes as possible. When her father remarried, his new wife was either unable or unwilling to cope and the children were sent back to their grandparents. Violet was then eight, couldn't read or write and hated having to wear more clothes. She proved to be too much for her grandmother too. Minnie had been governess to some children in Scotland before her marriage. It was felt she would be better able to deal with Violet; the brother Errol went to EAC and Minnie's sister.

Arriving at dates is a bit like fitting a jigsaw together. Violet's arm and part of her face appear in R's photo (12). As R was then 33 the date was 1896; Violet was nine at that time, having been born in 1887 (Rolls-Royce Motors sent a car to take her out on her 100th birthday in 1987). I think she came to live with Minnie and R no later than 1895, when she was eight; they were still at Chorlton-cum-Hardy. Violet was affectionate as well as high spirited and became known as 'Cuddles'. At first she was not allowed to help in the house; she was told to go and play with her doll - which she hated - and would skip instead. Later, at Brae Cottage (1898 to 1908, ages 11 to 21) she had the job of looking after the chickens, and keeping R's tobacco pouch filled; he smoked a pipe. She mentions how keen R was on his garden - lots of peaches and nectarines; there would have to be some end products for the labour put in! After her first effort at making pastry, R told her to go and get the axe from the cellar, so that the piecrust could be broken up and eaten! She comments on R's sense of humour and his fondness for children.

Minnie was very industrious and devoted her energies to running the household, writing letters, ordering seeds for the garden, and so on. She admired R very much and would take great care never to offend him. She had a horror of illness and associated with this a refusal to have children, undoubtedly too frightened of the various phases of pregnancies. She was a bad sailor too, as the Cape voyage showed. Violet, on the contrary, at one time wanted to be a nurse. My own impression of R's feelings for Minnie is that he valued her help, but his first love was always his work. There was obviously a very deep affection between him and Violet.

What really sunk the marriage was therefore the first, 1911, operation at Norwich. Vi says this was a bypass. I believe that Nurse Aubin was initially recruited only to see R through the period of recovery at Overstrand. CJ arranged for the rail and car journey to the Mediterranean in August and the winter cruise round it. The combination of seasickness and the inability to give nursing attention ruled Minnie out for this, so Nurst Aubin was kept on to cover it. She took full advantage of the opportunity. The only good thing Vi had to say about Nurse Aubin was that she was good at nursing. Minnie and Violet were virtually isolated from R at Crowborough, while R was recovering from his second operation, and again during a week at St. Margaret's Bay in 1914. Vi in particular, unaffected by illness, felt the exclusion keenly. Minnie and Vi remained at Crowborough for some time, until in fact they moved to Bexhill in 1921. Both had been much shaken by the move to West Wittering in 1917 and never saw Elmstead. Vi had a breakdown, and was still in hospital while Minnie went househunting in Bexhill. Minnie's sister also came to live at Bexhill after EAC died in 1921. Her marriage had been of much shorter duration than Minnie's and they had soon separated. R bought the house at Bexhill, 'Woodside' at Colenton Avenue, for Minnie and Vi; Minnie thought it was a present but on R's death found that it was still in R's name. R had ensured that she could continue to live there until she died and that she would be maintained. She survived until 1936.

What annoyed Vi most about Nurse Aubin was her extreme possessiveness and jealousy. Others also have commented on this. In my observation these are very often indications of a close physical relationship. There are reports too of Nurse Aubin's efforts to persuade R to divorce Minnie so that they could be married. CJ's briefing of Nurse Aubin was to maintain R in a state of health where he could continue to work for as long as possible. To achieve this Nurse Aubin would have to obtain a measure of control over R; the best method of getting this was up to her. It was also then very much in her own interest to keep R alive and well for as long as possible. She was two years younger than Vi; she enjoyed her position of power; and she also enjoyed acting as hostess to R's visitors, some of them well known and distinguished. Vi thought R was mesmerised by her, but he did maintain contacts with Minnie and Vi, at Crowborough and Bexhill. There were two visits to Vi alone, after her marriage. The last was when her two children, Hugh and Mary, were respectively three and one. This may have been the occasion when R had had an endurance trial car diverted on its way back from Chateauroux so that he could try it out. The trial consisted of a run to Bexhill, when the driver was instructed to leave R for a time and collect him later. R had also given Vi a generous present of £100 on her marriage. Vi was bitter too about intercepted letters from R, by 'that wicked woman'.

Nurse Aubin was therefore only about 45 when R died, and left her without a job. There were comments about her romance with Mr. Tyldesley from Claremont Haynes, the company's and R's solicitors, who dealt with R's affairs after his death. The marriage however was very short lived.

Vi referred to visits from CSR, at Quarndon, usually to tea ("Lots of thick bread and butter, Vi"). She thought he was a charmer, and very much the apple of his mother's eye. CSR was a good deal younger than R. He was born on 27th August 1877 and so was not quite 27 when the two first met, and only therefore about nine years older than Vi. He had graduated B.A. from Cambridge in 1898. Although he had first become interested in motor cars in 1896 he did not manage to buy one until two years later. He soon became interested in competitions and was on the Draft Rules Committee of the ACGBI for the 1900 "1000 Miles Trial", so capably organised by CJ. C.S.Rolls & Co. was only just starting out in 1902. It is easy to understand CSR being in awe of a mature and assured engineer of 41 when they first met, especially as he had produced a most impressive 10hp car and responded easily and quickly to the idea of producing progressively larger cars. CSR was neither a designer or innovator, so it was understandable that he considered R "the greatest motor car designer in the world" and that he appeared diffident in writing to the great man in 1909, especially as R had by then brought out the 40/50 'Silver Ghost'. The letter, written from Conduit Street, is reproduced in full in R-REC Bulletin No. 156. A few short extracts are interesting: "..I am going to look up the weights Wright gave me...to see exactly what we should have to do in the way of engines for his machines...I have got a Wright plane on order (private at present), if I could get one of our engines it would be a good ad... but I am afraid it would not pay us to do anything specially...I love your letters, they are so natural and everything is thrashed out so fully...I have done quite a lot of outside work for RR these few months - country house visits have been most productive and it is quite fun at the end of each visit to see and hear the whole house converted to RR cars and able to talk of nothing else...You know your designing has made such a reputation for the RR that one could hardly believe it could have been done in such a few years....".

CJ did stay with the Royces on occasions; Vi was fascinated by the way he would eat a number of boiled eggs in succession, with no accompanying toast.

In Chapter 9, I referred to CJ's approaches to possible successors to R - in case of need. R was not told of these; it would have been tactless if he had been. He did get an inkling from his Canadian cousin James Charles Royce which excited his curiosity. CJ was dead before W.O.Bentley was acquired as a part of the deal to get Bentley Motors. It is interesting to speculate on what his reaction to W.O. might have been if he had still been alive. R did meet W.O. once after the acquisition, in London early in 1932. R affected to believe that W.O. was only concerned with the commercial aspects of Bentley Motors, which could have been regarded as derogatory anyhow; W.O. was goaded into pointing out that he had had five years of apprenticeship with the Great Northern Railway, as compared with R's three. I think the 8 litre must have hurt R, and he couldn't be expected to welcome a young man of 43 who had achieved it only some nine years after his first car design; the Phantom II arrived twenty two years after R's first car design (admittedly minus four years of war). R at that time was 68.

The Rolls-Royce London office had acquired Bentley Motors primarily to kill the 8 litre Bentley. The use of the name on a Derby based design was secondary, the gilt on the gingerbread. In its last years Bentley Motors was financed by Woolf Barnato, who was in America when the slump broke in 1929, and no longer interested in Bentley Motors. His manager in England was unwilling to carry Bentleys through the slump and had already offered it to Rolls-Royce as a going concern before calling in the Receiver. It was turned down because Rolls-Royce thought the asking price was too high, but they will have known about W.O.'s contract with the company as a result. The purchase from the Receiver cost far less, still killed the 8 Litre and ham-strung W.O. Napier's had felt so certain of getting Bentley Motors that they had set up a design office at Acton and the design of a 4 litre engine for a Napier-Bentley was well advanced. That too could have worried Rolls-Royce.

R's rejection of W.O. meant that he was wasted at Rolls-Royce; he was only asked to see that the company's demonstrator cars were in good order. Hs did get him to take a 3.5 litre on a European tour to get his opinion of it. W.O. became early painfully aware of the front axle tramp on that model. He had met a similar problem on the 8 litre during development, and had found a cure. There was a mutual respect between Hs and W.O., who had first met in 1914 when W.O. was "selling" aluminium pistons.

So, if CJ had still been alive in 1932 and continuing his previous policy, there might very well have been a job for W.O. at Derby, perhaps occupying a role in relation to cars similar to Rg's on the aero side. I wonder what E's reaction to that would have been? He had only just arrived in Derby himself. There is no doubt in my mind that Hs would have backed W.O. rather than E in case of need. Idle speculation!

Another of Vi's comments was that R took Minnie and her out in experimental cars; she says they were the first passengers in the three cylinder, and remembers rides in the four and six afterwards. Their nickname for the 10hp twin that R used for a time was 'Mister Seven' (the registration number was NMR7). She also remembers a trip to Scotland in a test body, with no windscreen and only two bucket seats, and an egg box for Vi. It rained and they only had umbrellas for shelter. The trip was to visit Minnie's former employers near Moffat. On the way back they arrived at an Ullswater Hotel at 8.30 pm when dinner was over. The other guests, in evening dress, looked down their noses at their rather bedraggled appearance. On the following morning, when the guests had discovered who R was, they had quite a send off.

12. **Conclusion**

R died on 22nd April 1933, soon after reaching his 70th birthday. In "The Dean's Watch" Elizabeth Goudge wrote: "In fearing lest they faced disintegration without him men had forgotten the momentum that men of genius leave behind them in their works".

In this particular case the momentum acted in two ways. First it manifested itself in the products of the company, the cars and aero engines. These were the results of the product engineering history of the company. The development process and the organisation which embodied it would continue to roll forwards. Secondly, the whole corporate entity of Rolls-Royce, not only the engineering development department, had its own momentum, not confined to the few examples of staff fairly close to R who have been mentioned in this booklet but embracing all the different departments and their employees. It would indeed be far more difficult to stop the continuing progress than to let it continue. The simile suggests a pyramid, not firmly anchored on the ground but supported by a layer of Phantoms, 20/25s and 3.5 litres, all pointing more or less forwards, though some drivers might be a little wayward, with another layer of propellered Kestrels and Buzzards above them, also all reasonably aligned. How interesting that the same organisation had produced vehicles noted for their silence and engines in which silence was the last consideration.

APPENDIX: SIX CYLINDER CRANKSHAFTS

As will have been clear from the text, R was under considerable pressure from the time he met CSR in May 1904 until the 40/50 (Silver Ghost) was in production early in 1907. The vibrations of the out-of-balance Decauville had made him very conscious of the need for smoothness. The end web balance weights of his original twin were the economical way of counteracting all the rotating and some of the reciprocating weights. The amount of this counterbalance was decided before the car ran; it was smooth and quiet from the beginning.

For the three and four cylinder versions R was able to draw on existing practice as both types were fairly common at the time and CSR would have been familiar with their layouts.

The middle to end of 1904 however was too early to find much to study or copy in six cylinder engines. As mentioned in the main text, the Spyker of 1902 was the first six cylinder and Napier's first six, the L49, is dated 1903 - 1907 in "The Complete Catalogue of British Cars" (Culshaw & Horrobin, 1974, Macmillan). Other information however says that though it was announced in 1903 it first appeared early in 1904. Early sixes had three twin cylinder blocks; some, probably most, had 2+2+2 crankshafts. The unacceptable roughness of the first 30hp Rolls-Royce six makes it certain that R had this arrangement and this is confirmed by the eyewitness statement that it had "lots of balance weights". Let us look at the two options for such a crank; the superficially attractive one is crank 'A', which, as nearly as possible, puts succeeding crank throws opposite to each other, though at every gap between blocks the adjacent crankpins are at 120° to each other rather than 180°. The quick and easy solution would accept this arrangement and retain all the existing twin cylinder balance weights on the end webs of each twin cylinder component. Although there is 120° between adjacent crank throws at transitions between blocks, the adjacent twin cylinder cranks are at 60° to each other.

The alternative crank arrangement is that of 'B' (A1) in which succeeding twin cylinder crankshafts are at 120°, though adjacent throws are at 60° at inter-twin transitions. Because this consists of two three cylinder cranks at 180° to each other, staggered by one cylinder spacing, the primary couples cancel out; balance weights therefore could do no good.

To understand these out-of-balance couples we need to start from scratch with a single cylinder unit. There are two lots of weights to be considered; the rotating and the reciprocating. The former comprise crank webs, crank pins, the big ends and the lower part of the shanks of the connecting rods. The latter consist mostly of the cast iron pistons and the upper parts of the connecting rods. Rotating parts can be completely counteracted by rotating balance weights. For an in-line engine the inertia forces and couples due to the reciprocating weights are then confined to the plane of the cylinder centre lines. The basis for the forces is the product of the reciprocating weight and the crank radius. The primary forces during crank rotation depend on this product multiplied by the cosine of the angle of crank rotation from its top dead centre position; the secondary forces depend on that product multiplied by the cosine of twice the crank rotational angle and also by the ratio of the crank radius to the connecting rod length. Both also vary as the square of the rotational speed. For comparative purposes between different engines of the same bore and stroke, but different numbers of cylinders, one can ignore the speed effect; it will apply equally to all.

The formula for the inertia force is

$$\frac{Wr\omega^2}{g}\{\cos\theta + \frac{r}{L}\cos 2\theta\}$$

where W is the reciprocating weight, r the crank radius, L the connecting rod length, ω the angular velocity in rad/s, g the acceleration due to gravity and θ the angle of crank rotation of the crank from top dead centre. Because $\omega = \frac{2\pi N}{60}$, where N is the engine speed in rpm, the formula can also be written as

$$\frac{4Wr\pi^2N^2}{3600g}\{\cos\theta + \frac{r}{L}\cos 2\theta\} \qquad \text{which simplifies to} \qquad (2)$$

$$\frac{39.48WrN^2}{3600g}\{\cos\theta + \frac{r}{L}\cos 2\theta\}$$

For a twin with 180° crank the primary forces cancel out but the secondary forces are additive. Because the primary forces operate at the separation of the cylinder centres there are primary couples; there are no secondary couples. On 180° twins, and four cylinder in line engines the secondary forces have to be endured.

At least the rotating weights must be counterbalanced. The unbalanced reciprocating weight couples are then the base from which improvements could begin (A2). For calculation purposes the reciprocating weight is taken as 2lb, the crank radius is 2.5″, L is taken as 12.5″ (r/L = 0.2) and the cylinder centres for the twin as 5″, with increases for the three between adjacent cylinders, and for the six between the cylinder blocks, to allow for casting requirements. For the twin and three cylinder we know the crankpin arrangements, at 180° for the twin, at 120° for the three cylinder. I have already shown the two possible 2+2+2 crank arrangements for a six. Both have firing intervals of 120°-120°-180°-120°-120°-60°.

It must be appreciated that the unbalanced couples of (A2) are in the plane of the cylinders. If we attempt to counterbalance the reciprocating weight of an individual cylinder by rotating counterweights, only the components in the cylinder plane of those counterweights are therefore effective in reducing the reciprocating weight out-of-balance forces; the components across the engine are an unwanted by-product. For each cylinder therefore, and for each rotational position of the engine, we will have a resultant force in the cylinder plane, another transversely, and they combine to produce a resultant force at a specific angle. For a complete crank rotation the resultant forces at successive rotational positions can be shown as vectors and they produce a polar diagram.

For a multi-cylinder engine the positional shares of the resultant cylinder plane inertia forces and their associated transverse forces can be referred to the two end bearings of the crank. For the engines we are considering the resultants of the two, at each rotational position and at each end bearing are equal and opposite. One resultant force, at each rotational position, can be multiplied by the bearing separation to produce the out of balance couple. These also can be shown as vectors to produce a polar diagram for the couples. The directions shown are for the ones at no. 1 main bearing.

For the twin, with primary couples only, the curve of (A2) is a plain cosine one. The polar diagram, for 50% counterbalance, the ideal (A3), is circumscribed by a circle, with each vector half the maximum couple of (A2). For comparative purposes the values of (A2) are also shown on the left, as a single vertical line polar diagram.

For the three cylinder, although both primary and secondary forces cancel out, there are both primary and secondary couples and it is their joint existence which produces the distortion of the curve of (A3) and therefore of its associated polar diagram, again for 50% counterbalance of the reciprocating weights. By inspection, a counterbalance of rather more than 50% would produce the smallest maximum vector. In another three cylinder engine I investigated about 55% counterbalance gave the best compromise.

It is in the curves of (A2) and the polar diagrams of (A4) that we appreciate the major differences between the two six cylinder cranks. Both cranks consist of two three cylinder cranks staggered by one cylinder spacing, but the difference is that in A the angle between corresponding cranks of successive cylinder groups is 60°, and in B 120°. In both cases primary and secondary forces cancel out. In crank A there are considerable resultant primary and secondary couples and the spacing results in a figure of eight circumscribing envelope for the vectors. Equalising the two loops of the figure eight requires between 75% and 100% counterbalance of the reciprocating weights and even then the maximum vector is still about 1.5 times that of crank B. The advantage of crank B is that because it consists of two three cylinder cranks with 120° pin spacing in each, and one at 180° rotationally to the other, the primary couples cancel out and only the secondary couples remain. The other result of the primary couples cancelling out is that there is no benefit from counterbalancing the reciprocating weights. Whatever counterbalancing might be used, the out-of-balance couples are those shown in (A2) and the vertical line in the top right hand corner of (A4).

I have no direct evidence as to which crank configuration R chose. I rather suspect it was crank A. Bill Morton has confirmed that the sample he saw had a lot of counterbalance weights; crank A would have six. He has also said that the early sixes were unacceptably rough, which does suggest crank A with the initial use of the same balance weights as on the twin.

These early sixes did have timing gear rattle. R might have supposed this to be due to the slightly irregular firing order; a small front end flywheel would be a natural reaction to this. He tried various sizes and got some improvement but the larger ones led to broken cranks. One of the flywheels was first fitted by means of a wooden adapter, to save time. In this way it was more effective than when a metal adapter was substituted. R noticed marks of burning on the clamping faces of the wooden adapter. He deduced from this that relative movement had occurred, and that this and the friction between the clamping faces had been responsible for the improvement. He had discovered by chance the principle of the Lanchester crankshaft damper some years before Dr.F.W.Lanchester patented it. He did not immediately use it.

The photo of Albert Wheatman machining a six cylinder crankshaft (67), late 1904 or early 1905, shows why R did not need to use it. The solution to his problem was staring him in the face (A5). The four cylinder engine had cancelled out the couples of the twins by putting two twin cylinder cranks back to back. A similar solution could be used for the six, with two three cylinder cranks back to back in mirror image form. The very considerable advantage of this is that it gives an engine free of reciprocating weight forces and couples without counterbalancing. That this crank configuration was a feature of the six cylinder engine is clear from (67). R would have liked to convert all the sixes but may not have been able to. If any do remain it would be very interesting to find what form they took. Later six cylinder cranks had bearing sizes put up to 1.75″ diameter.

To complete the picture we need to have an understanding of the reasons for torsional vibrations in six cylinder cranks. The best known authority on these is Dr. W.Ker Wilson ('Practical Solution of Torsional Vibration Problems', Chapman & Hall, 1948, and also Proc.I.A.E. Vol.XXX, 1935-36, pp 453-527). Most of us are familar with the idea of the torsional pendulum (A6) because it is used in some ornamental clocks; it consists of a weight of torsional inertia J connected to 'earth' by a torsionally flexible member, which can be represented as a length L of a diameter d of steel. It is well known that, because the torque produced by this torsion spring is directly proportional to its angular displacement, the torsional vibration resulting from an initial angular displacement is of sinusoidal form and its frequency is independent of the amplitude. To maintain a given frequency it will be appreciated that L would have to be reduced in inverse proportion to any increase in J. If therefore we had a system such as that of (A7), in which an initial displacement of J_1 is produced in relation to J_2, the rotating masses will vibrate in opposite phase and the frequency will be appropriate to lengths of torsional spring L_1 and L_2 such that $J_1L_1 = J_2L_2$. The node or point of no deflection can be found from this. The natural frequency of each,

$$F = \frac{1}{2\pi}\left(\frac{I_pG}{JL}\right)^{\frac{1}{2}} \tag{3}$$

where $I_p = \frac{\pi d^4}{32}$, d being the diameter of the torsional spring in inches, G = shear modulus of the material = 12×10^6 for steel, J = polar moment of inertia of the disc, $\frac{Wk^2}{g}$, k being the radius of gyration, inches, and g the acceleration due to gravity = 386 in/sec^2, and W the weight of the disc, 1b. Equation (3) therefore becomes

$$F = \frac{1}{2\pi}\left(\frac{\pi d^4 \times 12 \times 10^6 \times 386}{32 \times Wk^2L}\right)^{\frac{1}{2}} \tag{4}$$

$$= 3.394 \times 10^3 \left(\frac{d^4}{Wk^2L}\right)^{\frac{1}{2}} \text{ Hz}$$

$$= 203.64 \times 10^3 \left(\frac{d^4}{Wk^2L}\right)^{\frac{1}{2}} \text{ per min}$$

The vibration system for a six cylinder engine can be represented by (A8), the individual inertias J_1, J_2, etc corresponding to the different cylinder units and J_7 to the flywheel. Dr.Ker Wilson has shown that the system can be reduced to a two inertia system for approximate calculation purposes, one being J_7 and the other the sum of the inertias J_1 to J_6. The torsional coupling between the different inertias is expressed in terms of length or equivalent length of the main and big end bearings and the webs, calculated according to formulae which are a mixture of basic theory with empirical modifications. For the approximate method just referred to the effective length is about half the total effective length of the crank as a whole.

The calculation of the inertias is straightforward but tedious (less so now than when I first made such calculations); arriving at the equivalent lengths of the components such as crank webs is more open to argument, especially when there is overlap between the main bearings and the crank pins. From the vibrating system so obtained frequencies can be obtained, an approximate one by the two inertia method referred to above, a more accurate one by tabulation and iteration.

For a vibration to build up there has to be an energy input; the two main sources of this are the three explosions per revolution of the in-line six cylinder four stroke engine, and the interchanges within the crankshaft length of the variations during the stroke of kinetic energy of the reciprocating weights, at twice per revolution for each weight. The latter is the more important; in the six cylinder engine the 3rd order vibration is the most destructive. If the frequency of the system is calculated as F_m/min then this third order occurs at F_m/3rpm. The amplitudes obtained depend on the amount of damping which exists or can be provided within the system, and the individual energy inputs. Unless the damper is very effective it is dangerous to get too close to the speed of this 3rd order.

To be concerned only with the 3rd order is to oversimplify the problem, as the diagram (A9), based on information in Dr. Wilson's I.A.E. paper, shows. One of the implications of that diagram, with at least three fairly important orders below the 3rd, is that the timing gear noise of the 30hp six would have been noticeable at several speeds in the range. For instance, with an assumed frequency of the basic system of 5700/min, the 6th, 4.5 and 3.5 orders would occur at 950, 1270 and 1630 rpm; minor ones might be heard at 1035, 1140 and 1425 rpm.

Adding a front flywheel would reduce the amplitude of the vibrations there, and hence the timing gear noise; it would also reduce the speeds at which the noise peaks occurred. The exception to the noise reduction would be that due to the 3rd major, where the calming effect of the flywheel would almost certainly be overweighted by the effect of the reduced natural frequency putting the maximum running speed further up the flank of the vibration.

Being able to throw away the balance weights when he adopted the new crank layout taught R a lesson which he applied very successfully in the 40/50. The moment of inertia of the rotating masses varies as the square of their radii; $(4.5/5)^2 = 0.81$. The stiffness of the bearings varies as the fourth power of their diameter; $(2.25/1.5)^4 = 5.06$. The increase in length would be about 6 (4.5 - 4) = 3″, some of which could be used to stiffen the webs. The pistons however would be heavier. Only a detailed calculation would show exactly how much the improvement was; it was in any case enough to cope until the stroke was put up to 4.75″. The slipper drive then provided the answer (110).

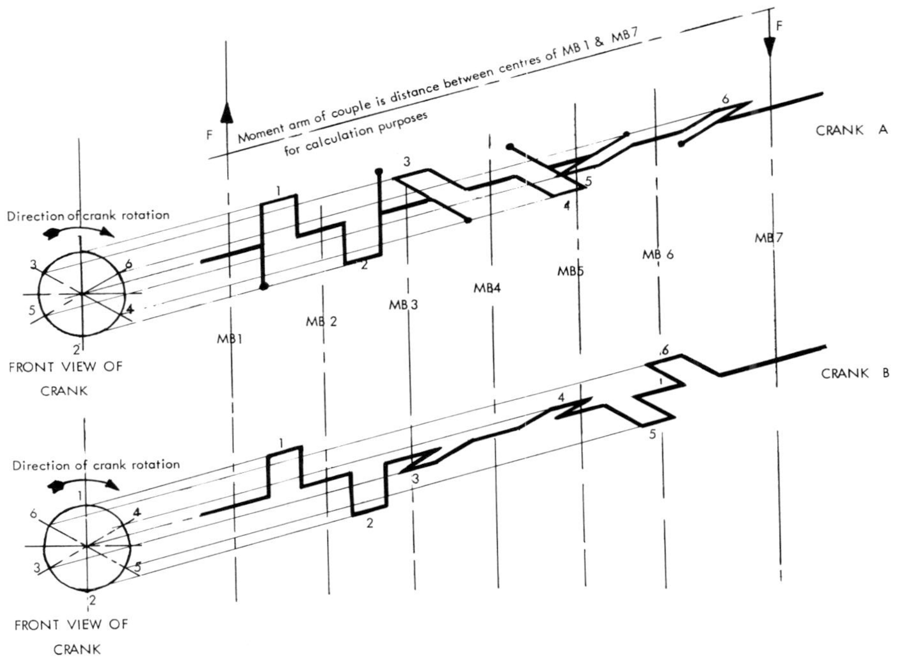

Figure A1 This has perspective views of cranks A and B for the six cylinder engine; for the early engines there was a choice between these two once the idea of the 2+2+2 crank was accepted. In crank A the balance weights shown are assumed to be as on the twin cranks (52); crank B does not require balance weights, not even for the rotating weights. In this respect it resembles the now universal crank form to which R fairly quickly changed over, which is inherently free from both primary and secondary forces and couples. The consequent elimination of the balance weights increased the torsional frequency of the shafts enough to avoid the breakage problems of the early sixes. The method of calculation refers the unbalanced forces at each crank to either the front or rear bearing. The resultants in cranks A and B are equal and opposite. One such resultant multiplied by the separation of the two bearings, as indicated on the drawing, is the resultant couple. The graphs show couples based on the direction of the resultant forces at no. 1 main bearing.

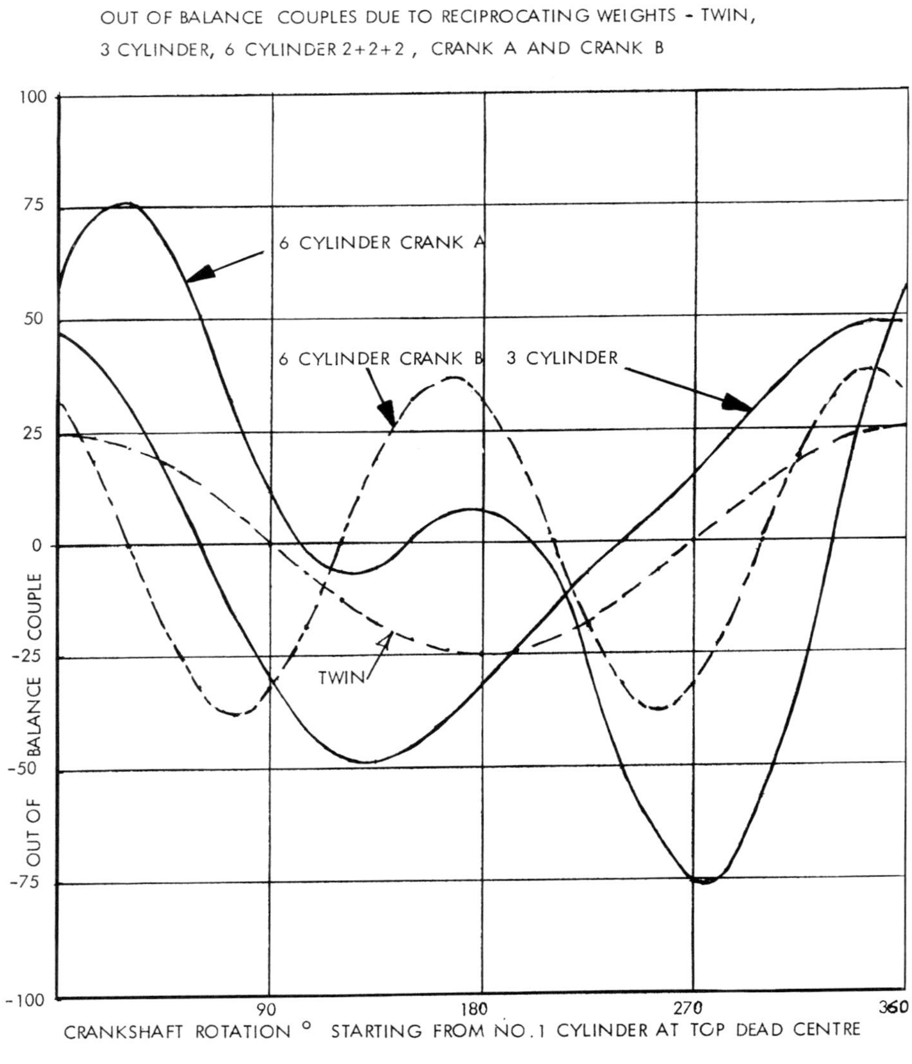

Figure A2 The unbalanced reciprocating weight couples for the twin, the three cylinder and for cranks A and B in the six cylinder, are shown here. The Decauville, completely unbalanced, would have been considerably worse than the Rolls-Royce twin and would have had transverse components too. The Rolls-Royce twin, with primary couples only, is a plain cosine curve. The three cylinder shows some distortion due to the secondary couples, crank A on the six cylinder a good deal more. With crank B there are secondary couples only and these occur at twice the frequency of the predominant primaries in the others. These unbalanced reciprocating weight couples act in the plane of the cylinder centres. They are plotted on a base of degrees of crank rotation from no.1 cylinder at top dead centre.

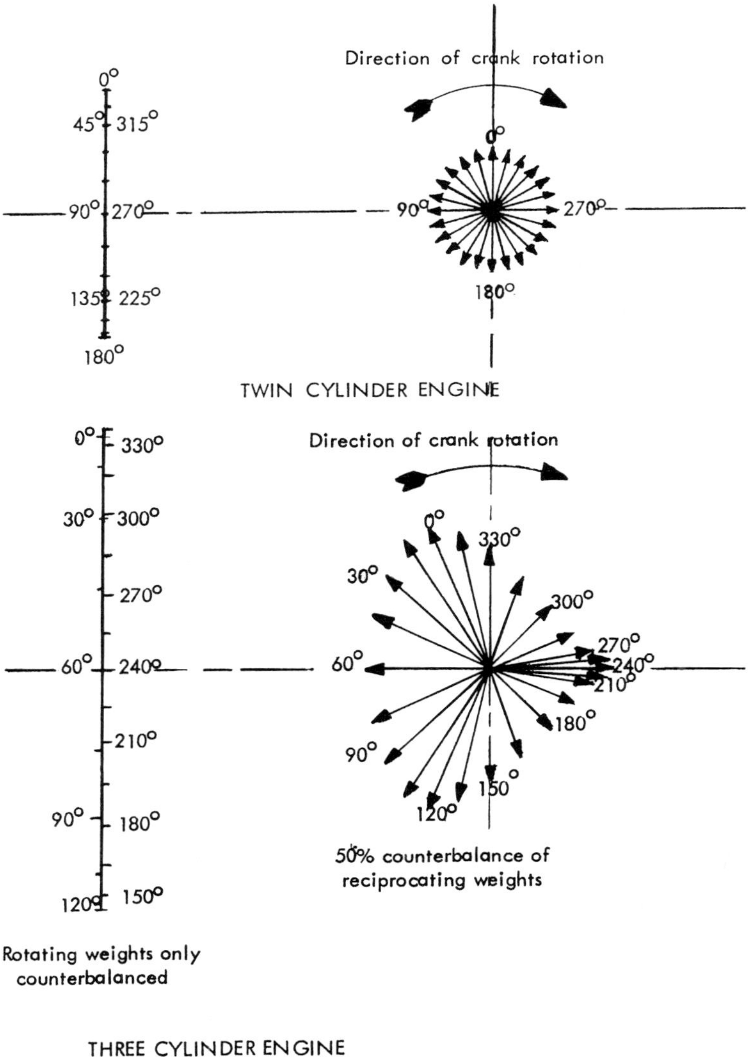

Figure A3 The unbalanced couples shown in (A2) are in the plane of the cylinder centres; they are represented by the single line 'polar' diagrams shown on the left, for the twin and three cylinder engines. Because the twin has only primary couples, counterbalancing 50% of the reciprocating weights will result in the polar diagram on the right with a circular circumscribing for the equal vectors, each of which is half the unbalanced couple shown on the left. The vectors are contra-rotating, as are also those of the three cylinder shown below; for these the envelope is slightly distorted by the secondary couples. Increasing the counterbalancing percentage spreads the envelope horizontally and squeezes it vertically; here rather more than 50% would be the ideal. Even so, the three would have a maximum couple about twice as much as the twin.

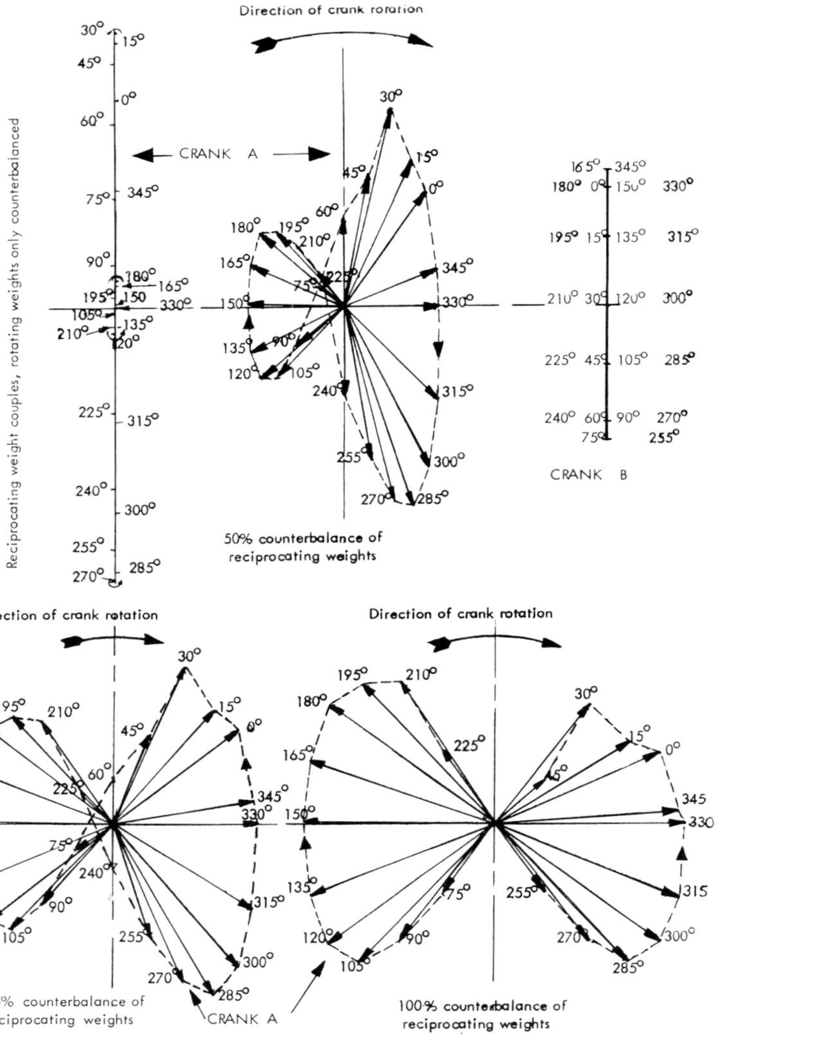

Figure A4 Reports of roughness on the early sixes suggest it was more likely to have had crank form A rather than B. Even B has maximum couples about three times as much as the 'ideal' twin, and therefore about 1.5 times as much as the 'ideal' three. With 50% counterbalance crank A has maximum couples about 1.5 times as much as those of crank B, and therefore about 4.5 times as bad as the twin. Study of the diagrams suggests that the 'ideal' crank A would need between 75% and 100% counterbalance. Even with the assumed 50% inherited from the twin the cranks were in torsional trouble. The torsional frequency lowered by the inertia of the balance weights brought the troublesome periods within or very close to the running range, resulting in the breakages experienced. The figure eight envelope of the vectors in the polar diagrams has contra-rotating couples in the right hand part of the diagram and forward rotating in the left.

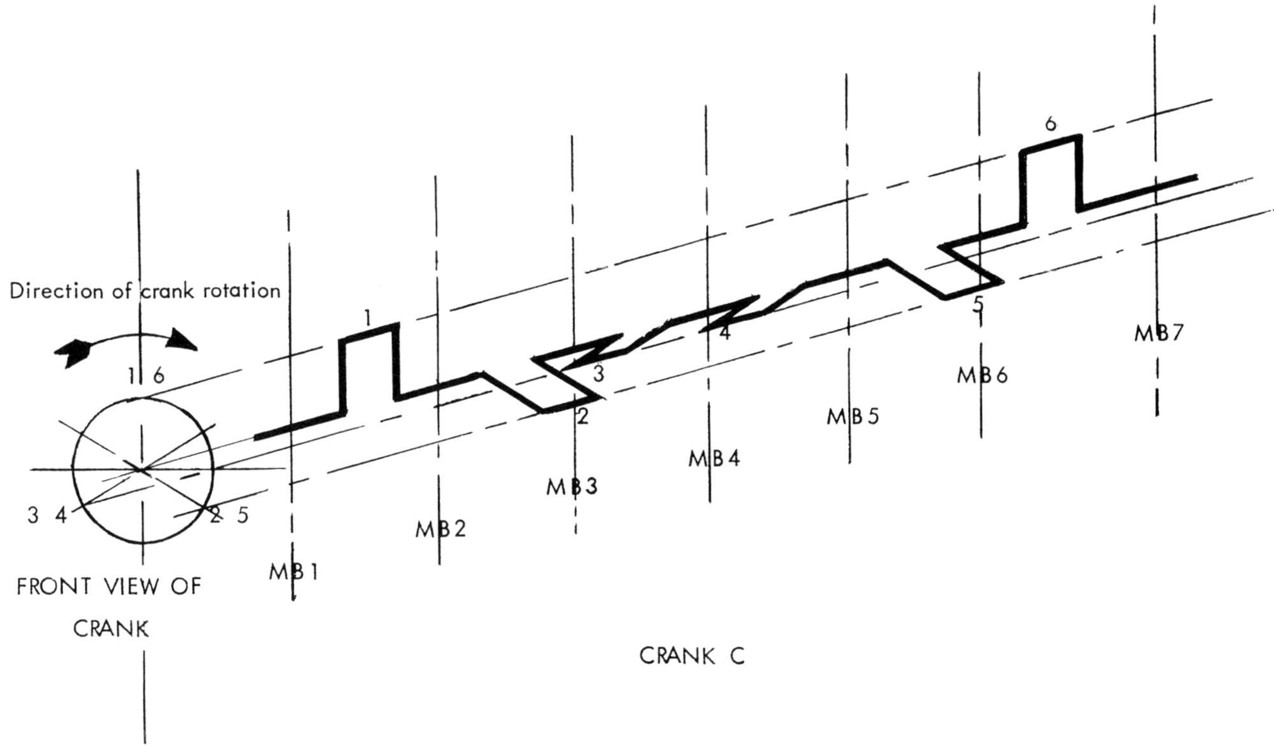

Figure A5 This is the form of crank which, with its mirror image, has become the standard for six cylinder engines. This, with the direction of rotation shown, gives a firing order of 1-4-2-6-3-5, for many years a feature of Rolls-Royce, and Bentley, cars. The mirror image crank, with cranks 2 and 5 where cranks 3 and 4 appear on the front view above, would have the firing order 1-5-3-6-2-4.

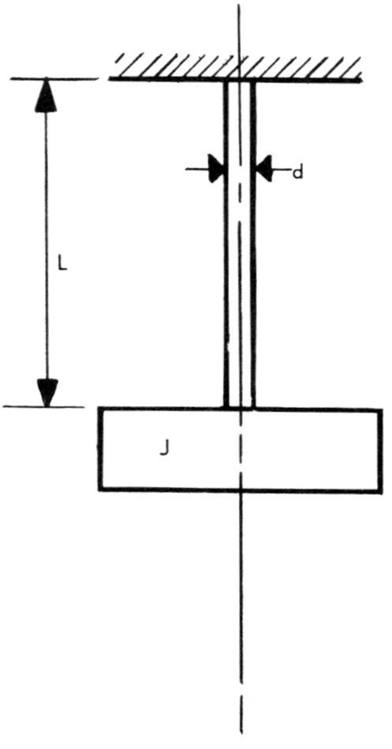

Figure A6 The torsional pendulum, familiar because of its use in some ornamental clocks.

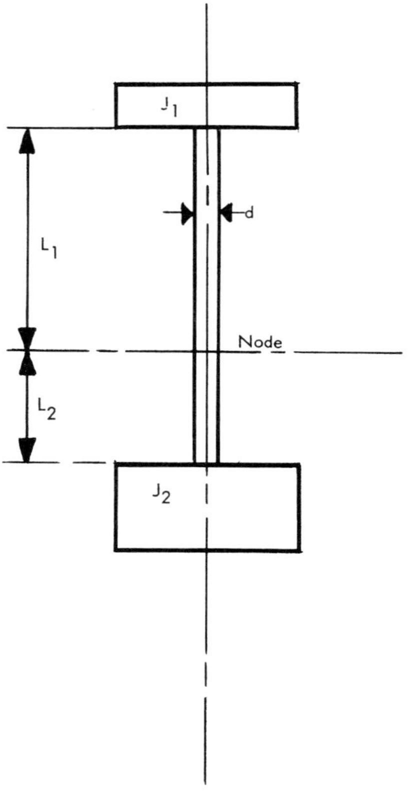

Figure A7 Two torsional pendulums back to back, the basis of the simplified vibrating system of the car engine crankshaft used for approximate frequency calculations.

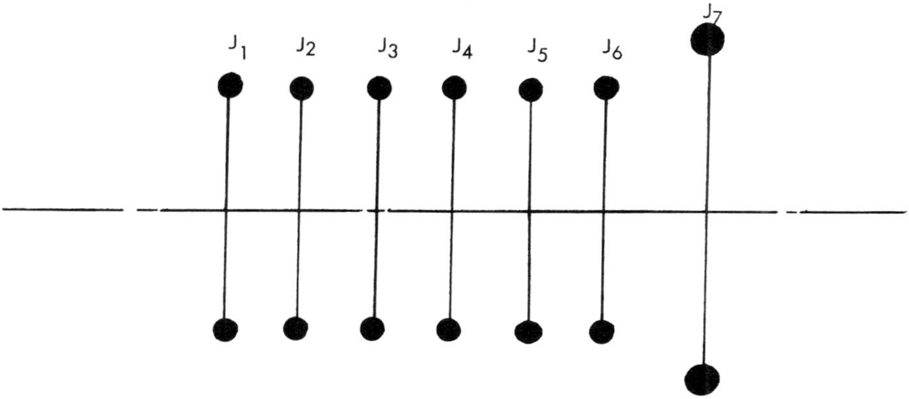

Figure A8 This is a representation of the actual vibrating system of a six cylinder car engine, assuming negligible inertia of the items fixed to the front end of the crankshaft, and without torsional vibration damper. Readers are referred to either of the quoted publications by Dr.Ker Wilson for details of the iterative tabulation system for arriving at the frequency of this arrangement.

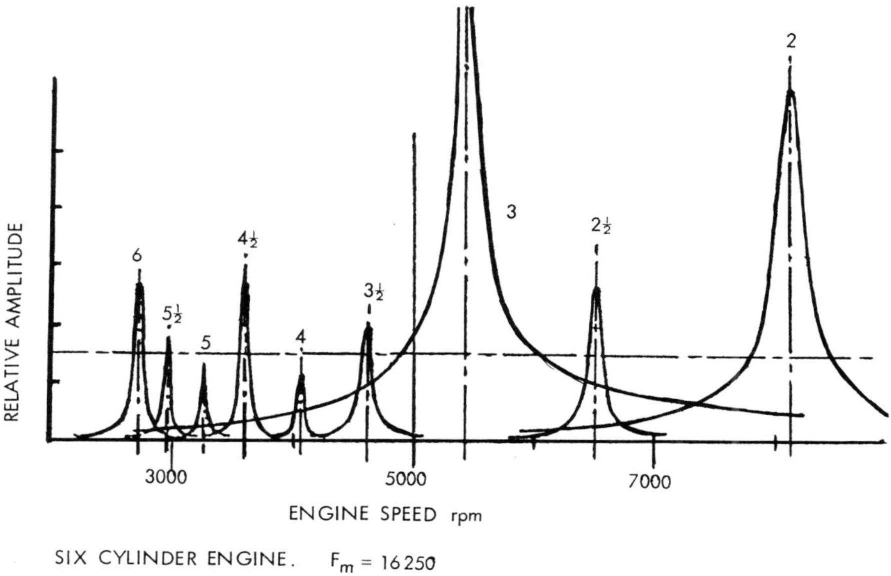

Figure A9 This diagram of the relative amplitudes of the different orders of vibration which occur in six cylinder car engines is based on information in the IAE paper by Dr.Ker Wilson. The vertical ordinates are an indication also of the maximum stress occurring at the crankshaft node, normally between inertias J_6 and J_7.

ROYCE FAMILY TREE

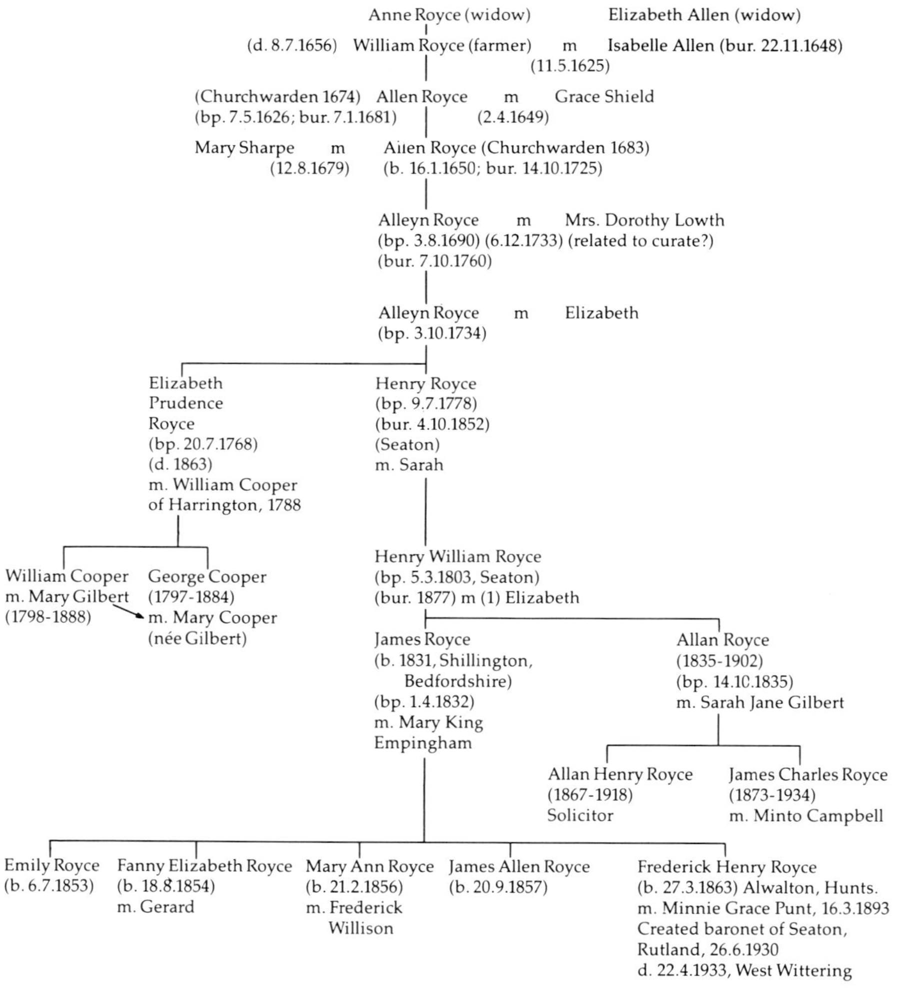

Figure 1 This simplified family tree incorporates the essentials, starting from Queen Elizabeth's reign, during which the first Royce entry, a widow, was born; earlier records may yet be found. Significant figures are Henry William Royce, his aunt Florence Elizabeth Cooper, two of her sons, William and George, R's uncle Allan and two of his sons, R's Canadian cousins who crop up later in the story.

Figure 2 Henry William Royce (1803-1877) of Seaton, R's grandfather.

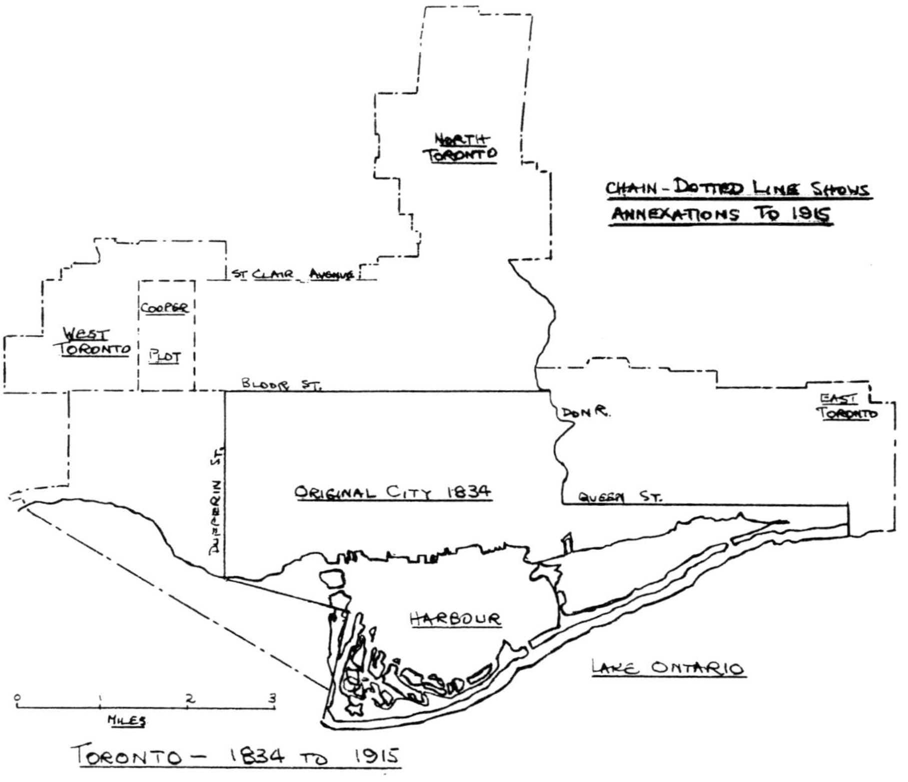

Figure 3 The site of the George Cooper 400 acre plot in relation to the 1834 Toronto boundary (full line) and the enlarged one of 1915 (dotted line).

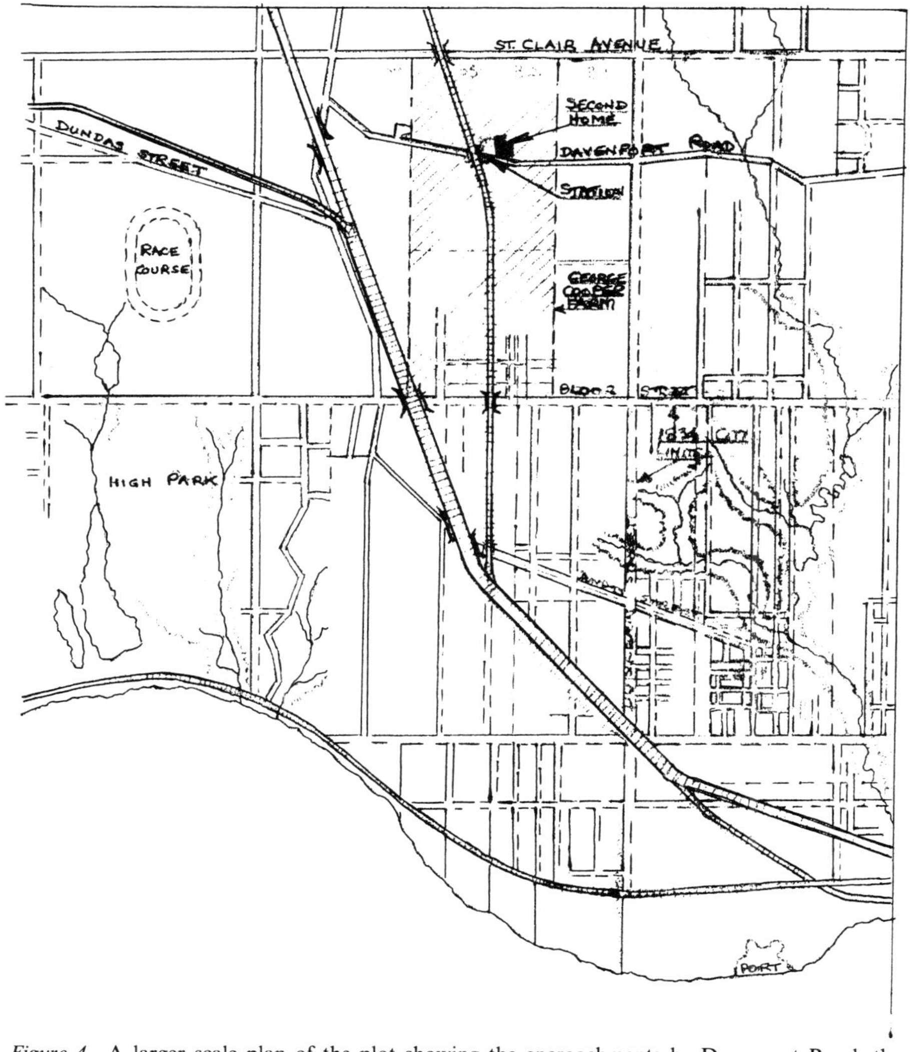

Figure 4 A larger scale plan of the plot showing the encroachments by Davenport Road, the local railway line and its Davenport Road station, and the main line railways which reached Toronto in the middle to late 1850s. The original George Cooper farmhouse, and its stabling for 50 horses, was in the southwest corner of the plot, but when the trunk railways arrived a new house was built which overlooked Davenport Road station. Other members of the family settled in parts of the plot, and it was gradually built on too.

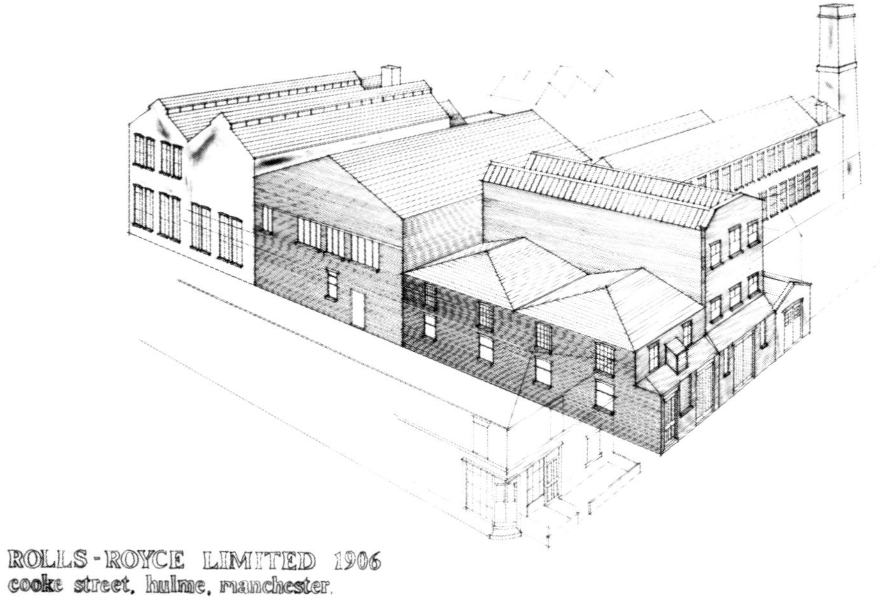

Figure 5 A perspective view of the final extent of the Cooke Street premises, in isolation, showing how much they grew from the original occupation of No. 1A.

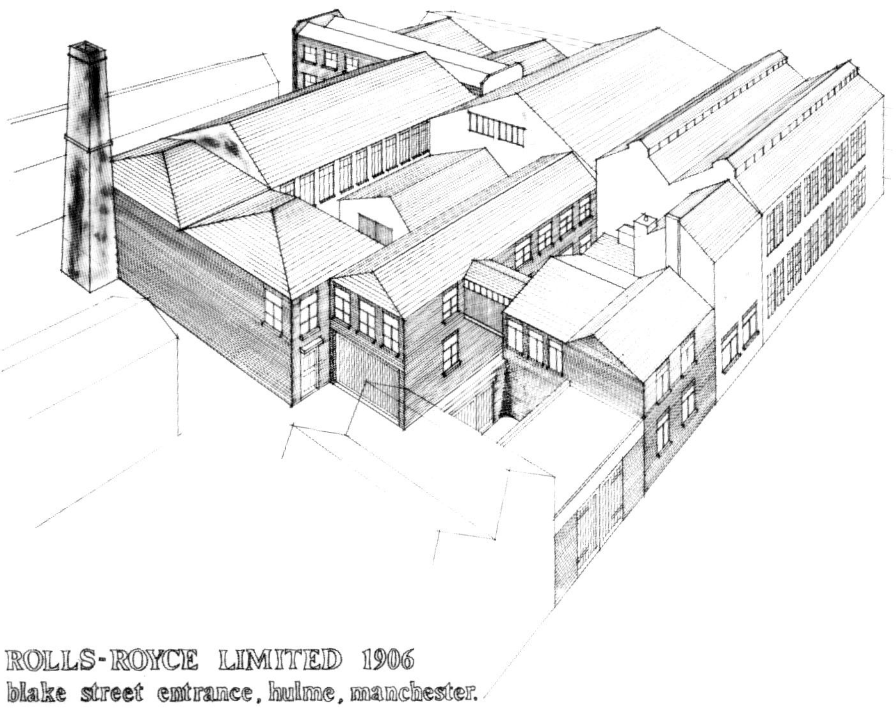

Figure 6 The corresponding perspective view of the Cooke Street premises from the opposite direction. Royce Ltd. finally burst out from here in 1899 or 1900. It continued to be very busy and was the source of the car business of Rolls-Royce. All the early cars were made here: the 10hp twin based models, the short-lived landaulet "Town Carriage" and "Legalimit" using the V-8 engine, and the early 40/50s including the original "Silver Ghost". Car production was moved to Derby in 1908, but some activities remained at Cooke Street until 1910, when the site was finally deserted.

Figure 7 The Cooke Street timing clock after transfer to Derby.

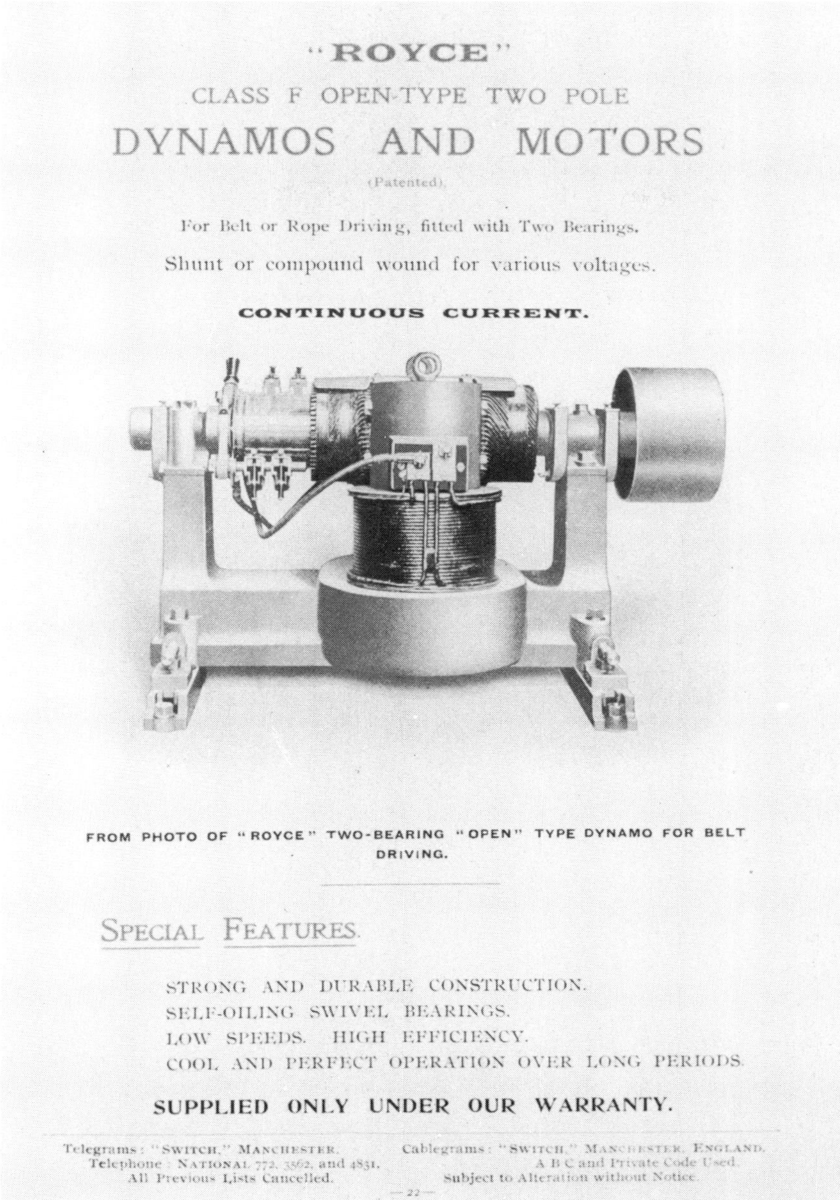

"ROYCE"

CLASS F OPEN-TYPE TWO POLE

DYNAMOS AND MOTORS

(Patented).

For Belt or Rope Driving, fitted with Two Bearings.

Shunt or compound wound for various voltages.

CONTINUOUS CURRENT.

FROM PHOTO OF "ROYCE" TWO-BEARING "OPEN" TYPE DYNAMO FOR BELT DRIVING.

Special Features.

STRONG AND DURABLE CONSTRUCTION.
SELF-OILING SWIVEL BEARINGS.
LOW SPEEDS. HIGH EFFICIENCY.
COOL AND PERFECT OPERATION OVER LONG PERIODS.

SUPPLIED ONLY UNDER OUR WARRANTY.

Telegrams: "Switch," Manchester.
Telephone: National 772, 3362, and 4831.
All Previous Lists Cancelled.

Cablegrams: "Switch," Manchester, England.
A B C and Private Code Used.
Subject to Alteration without Notice.

— 22 —

Figure 8 Plan of open type two-pole dynamo/motor. The field coils, commutator and brushes are prominent.

Figure 9 Multi-polar open type motor, with control and metering equipment.

Figure 10 The plate on this 10hp motor implies an efficiency of 89%.

Figure 11 Open type multi-polar dynamo of considerably greater power; the brush gear in the foreground implies a large current passing. The 1904 catalogue lists sizes of such dynamos up to 595 input horsepower, to produce 420 kw at 440 or 550 volts.

Figure 12 R at the age of 33, vigorous and in his prime; the hand, arm and part face belong to his adopted daughter, Violet Punt ("Cuddles"). The date, by inference, is 1896.

Figure 13 Ernest A.Claremont (EAC), R's partner and brother-in-law who looked after the commercial side of their business. He became chairman of Rolls-Royce from its inception and remained so for many years. This photo dates from about 1907.

Figure 14 A part of one of the Cooke Street workshops.

Figure 15 An aerial view of the Trafford Park works – Royce Ltd. in near foreground and W. T. Glover beyond.

Figure 16 A street view of the front of the Trafford Park works. The factory on the left is that of W.T.Glover & Co. Ltd., the cable manufacturers, run by Henry Edmunds, called the "Godfather of Rolls-Royce" because he organised the meeting of the two men in the Midland Hotel at Manchester. He was a director of Royce Ltd. at the time.

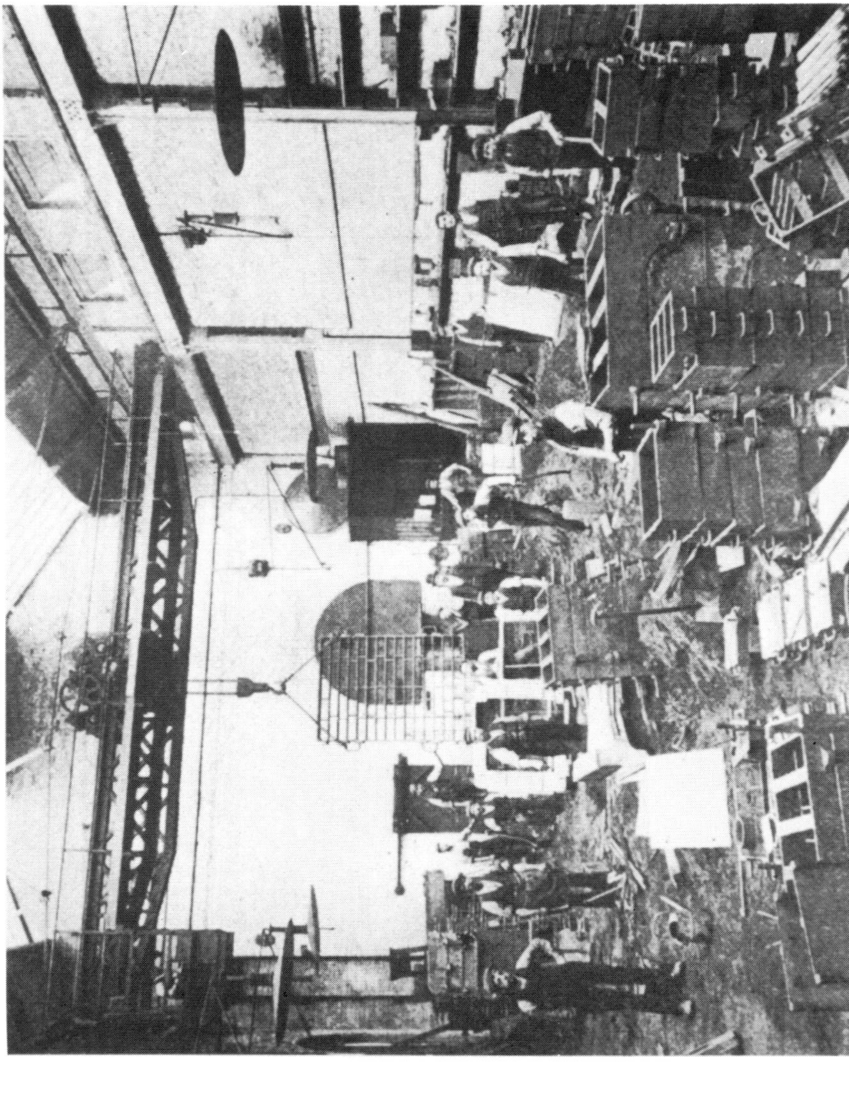

Figure 17 A gantry type crane in the Trafford Park workshops.

Figure 18 Another gantry type crane either in use or possibly under test at Trafford Park.

Figure 19 The Royce gantry type crane still in operation at Nightingale Road.

Figure 20 A closer view of the operator's cage and bridge structure of this crane.

Figure 21 Bomb damage in one of the side rails of this crane, from the early morning lone raider in World War II. From my home at Duffield I saw this plane come out of the clouds and return to them, getting its bearings minutes before it bombed the works.

Figure 22 Brae Cottage, Legh Road, Knutsford, built for R in 1898, five years after his marriage and fourteen after starting the business. Photograph by Rolls-Royce Motor Cars Ltd.

Figure 23 The panel on the gable end of Brae Cottage gives the date 1898, and the letters R for Royce, the reversed FH for him and MG for his wife.

Figure 24 A beautifully restored 10hp Decauville car in action; the present owner and restorer, John Welch, and his wife are on board.

Figure 25 A more restful view at John Welch's home.

Figure 26 The 3/4 front view of the car shows that not only the front axle but also the cross steering tube are bent down to clear the front of the engine. The convoluted finned tube forming the radiator would be inadequate without the rear mounted 7 gallon tank, and its connecting tubes.

Figure 27 The offside view of the engine shows the water outlet from the top of the cylinder block to the radiator, more detail of the radiator and its fins, and the automatic inlet valves.

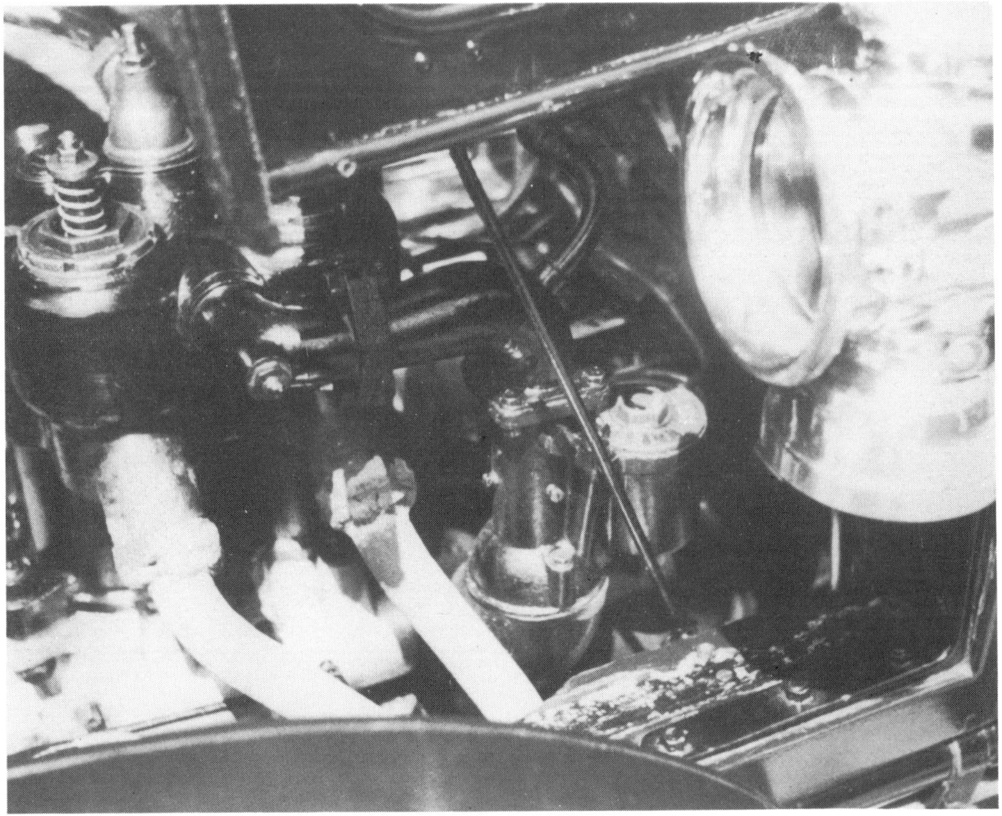

Figure 28 The nearside view shows the exhaust pipes, the carburetter position behind the entrance to the inlet passage cast in the cylinder block, a closer view of one of the inlet valves, and one of the oil headlamps.

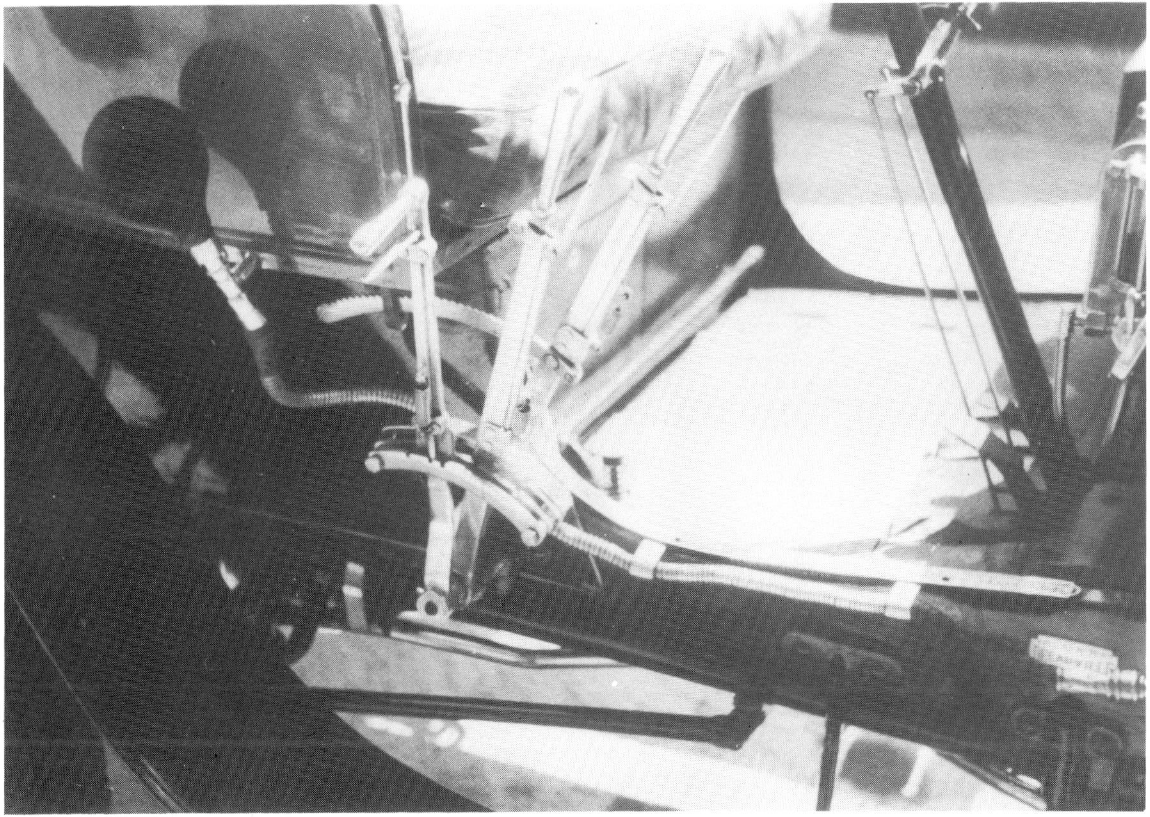

Figure 29 The two gear change levers, one for each selector rod, and the handbrake lever can be seen in this view. Copper piping took the cooling water from the base of the radiator to the rear mounted tank and back again to the pump on the engine which pushed it into the cylinder block.

Figure 30 The crankshaft, round shanked connecting rods and cast iron pistons are seen here out of the engine. Note the absence of any balance weights on the crank, and the two long bearings.

Figure 31 The crankshaft and one connecting rod are assembled into the crankcase lower half. Bearings were hardened steel on steel and the mains had ring oilers. The small gear on the right is for the camshaft drive; the large one drives the water pump. The upper half of the crankcase can be seen in the background.

Figure 32 Some cars, though not John Welch's, have a torque arm on the rear axle as shown here. At a guess, R's car probably did have this rather than the torque tube.

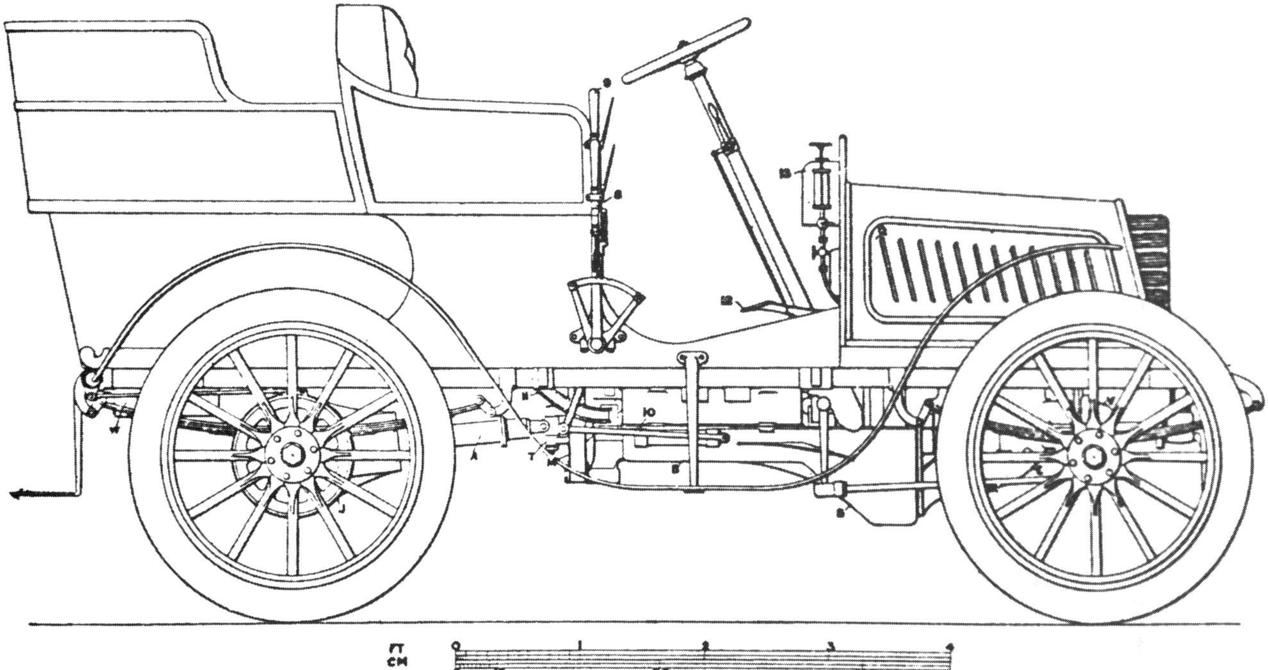

10 HP. Decauville Car, with Tonneau Body, 1902-3 Type.

Figure 33 A side elevation of the complete car has scales in feet and centimetres. The step for the rear mounting will be noticed, and the dashboard oil feed arrangements.

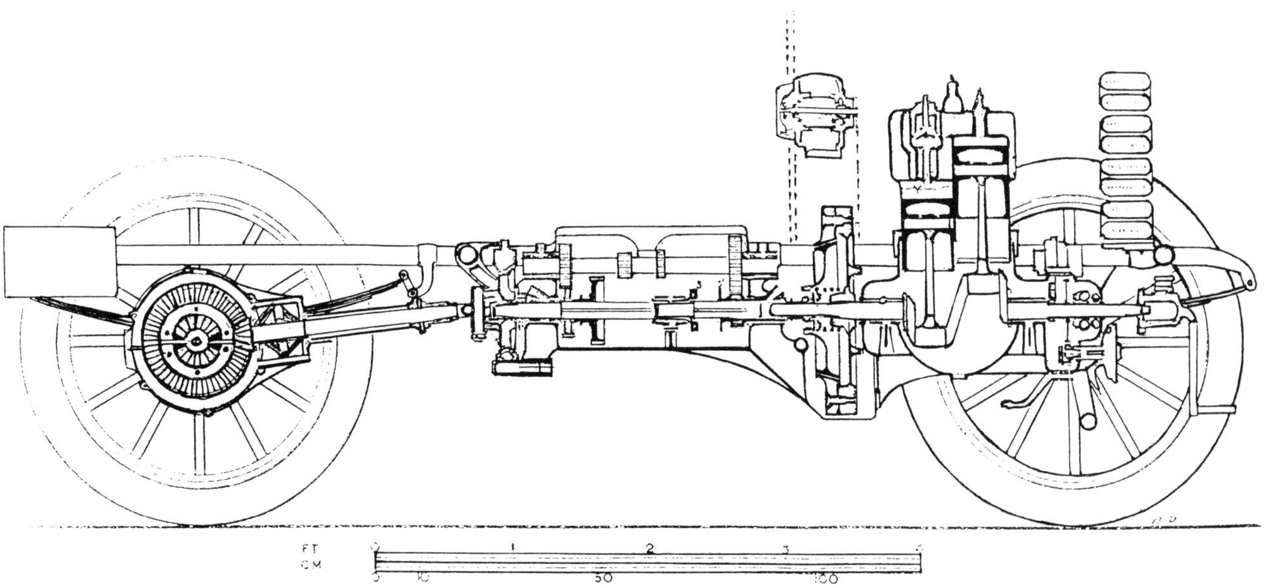

Figure 34 A sectioned side elevation of the chassis shows the torque tube rear axle, the unit construction of engine and gearbox, though with an open clutchpit for ready access, the large gearbox and the external contracting brake behind it. The Hooke's type universal joint behind the gearbox has offset centres to allow through pins to be used, and takes the axle torque forces in addition to its shaft torque-transmitting function.

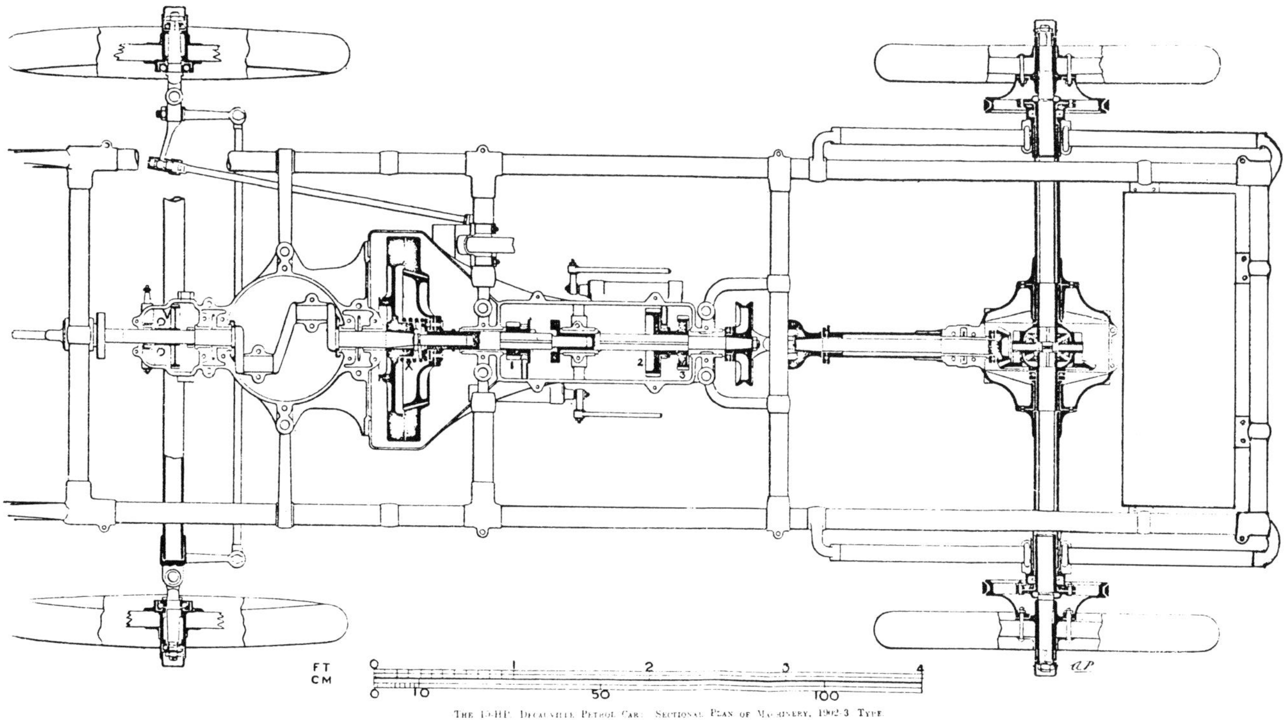

Figure 35 The sectional plan shows the tubular frame with brazed-on lugs for joints, straight parallel side rails with the rear springs offset outwards, the six rigid engine-gearbox unit mounting points and the external contracting (into vee grooves) rear wheel and transmission shaft brakes. The governor at the front of the engine was then normal.

Figure 36 How not to do it: R tried out on his Decauville an exposed chain driven overhead camshaft to operate the inlet valves, but would not have accepted such a large overhang between bearing and sprocket. This example was found some years ago and there was speculation that it might have been R's actual car. In this case the inlet pipe elbow is arranged to bring the carburetter to the front of the engine - compare with *figure 28*.

Figure 37 Henry Edmunds, who introduced Rolls to Royce in May 1904.

Figure 38 The photographic side view of the Royce 10hp twin shows the quadrant gear change lever alongside the hand brake lever, the channel section frame, with forged or malleable iron brackets, carburetter on the right hand side and behind the engine centre line, convenient for hand column controls, with a long vertical pipe so as not to prejudice mixture distribution, the oil feed on the dash, rear springs shackled at both ends.

Figure 39 The photographic plan shows the separate gearbox, track rod in front of the axle, away from the engine so that it can be straight, frame side members kinked to be wider at the back, torque arm on the axle, channel section side members, anti-lozenge diagonals across the rear corners, inlet valves over the bores and the original radiator top tank shape.

Figure 40 The front elevation of the Royce 10hp twin shows that the axle is tubular like that of the Decauville.

Figure 41 The gearbox casing is split horizontally at shaft level, with the layshaft alongside the input and output shafts. These latter are straddle mounted, the output through its spigot bearing in the input; there is dog engagement for direct drive and successive rearward movements of the dog-cum-sliding gears assembly on its splines by the single selector fork rod would then engage 2nd speed, 1st speed and reverse. The gear cluster here is between 2nd and 1st gear positions. The reverse idler can be seen behind the smallest gear at the right of the layshaft. With this arrangment reverse has to be lower than 1st gear; and it produces a long gearbox with long shaft spans between bearings, especially the layshaft.

Figure 42 Parts of the back axle centre assembly; note the straight bevel gears with straddle mounted pinion and the double gear clusters of the straight spur differential. Plain journal and ball thrust bearings were used.

Figure 43 The complete Royce 10hp car. Note forged front dumb irons, rear entry body.

Figure 44 The original Royce 10hp engine: gear type water pump at front of engine, governor housing below it. Longuemare carburetter, second camshaft for inlet valve push rod operation, and covers on block casting for core support and sand removal from water jackets.

Figure 45 The Hon. C.S.Rolls.

Figure 46 A Rolls-Royce 10hp car. The familiar radiator shape has now arrived.

Figure 47 This 10hp car, no. 20165, was donated back to Rolls-Royce after covering 100,000 miles. The spare tyre and rim carried ready inflated could be clamped to a wheel alongside a punctured tyre to get you to the nearest garage.

Figure 48 For the 15hp and larger cars "platform" rear springing was introduced: an inverted semi-elliptic spring across the back of the frame was connected to the rear ends of the fore and aft semi-elliptic springs to increase the deflections at the axle. The axle would be moved back to behind the middle of the fore and aft springs to achieve the right combination of axle up and down movement and axle attitude to match the requirements of the torque arm.

Figure 49 A 15hp three cylinder car; the longer wheelbase and wider track made possible better accommodation for the passangers, including in some cases closed bodies. This implied side entry, which also made hoods available when called for.

Figure 50 A 20hp car, the type known later as the 'Heavy 20' to distinguish it from the lightened version prepared for the T.T. races in 1905 and 1906.

Figure 51 The 30hp car with enclosed bodywork and noticeably forward radiator position for the time. R had shortened the front of the multi-cylinder engines, figures 57 and 61, by the revised auxiliary drive arrangement.

Figure 52 For the Rolls-Royce 10hp twin a centre main bearing was added (without increasing the cylinder centres) to reduce the deflections of the front end of the crankshaft. The bearings are still in the crankcase lower half, a carry-over from stationary engine and Decauville practice. Note the counterbalance weights on the end webs. By eye they balance out the whole of the rotating weights (pins and big ends) and a proportion of the reciprocating weights too.

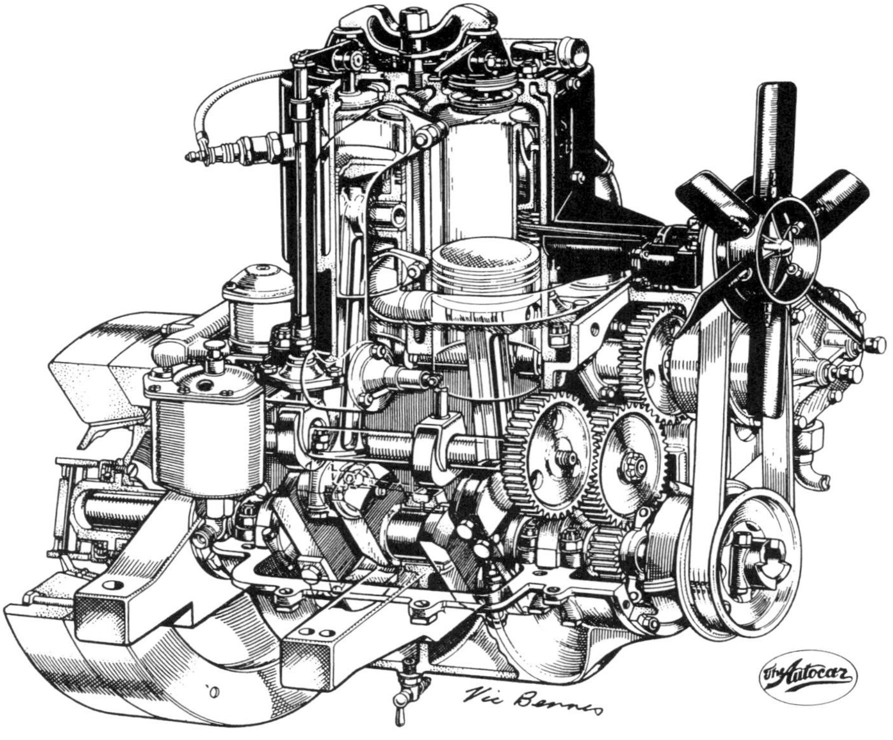

Figure 53 This Vic Berris drawing from the "Autocar" is of the Rolls-Royce 10hp twin, with separate camshafts for inlet and exhaust valves. There is a radiator fan driven by a flat belt, and by now a Royce carburetter.

Figure 54 The nearside of the 15hp three cylinder engine. The governor is on the vertical shaft inside the conical housing. Note the shaft drive to the ignition distributor and the pump for the separate oil feeds, the expansion chamber at the exhaust ports which persisted in various forms until the "New Phantom" engine, and with the individual cylinder castings, the inlet valves over the exhaust. There was only room for the water outlet, inlet pipe and inlet valve at the top of each cylinder if the inlet valve housing could replace the cap which had to be above the exhaust valve for assembly. To meet this inlet valve position a single camshaft now operates both inlet and exhaust valves, an arrangement which carried over to the larger engines though not essential for them.

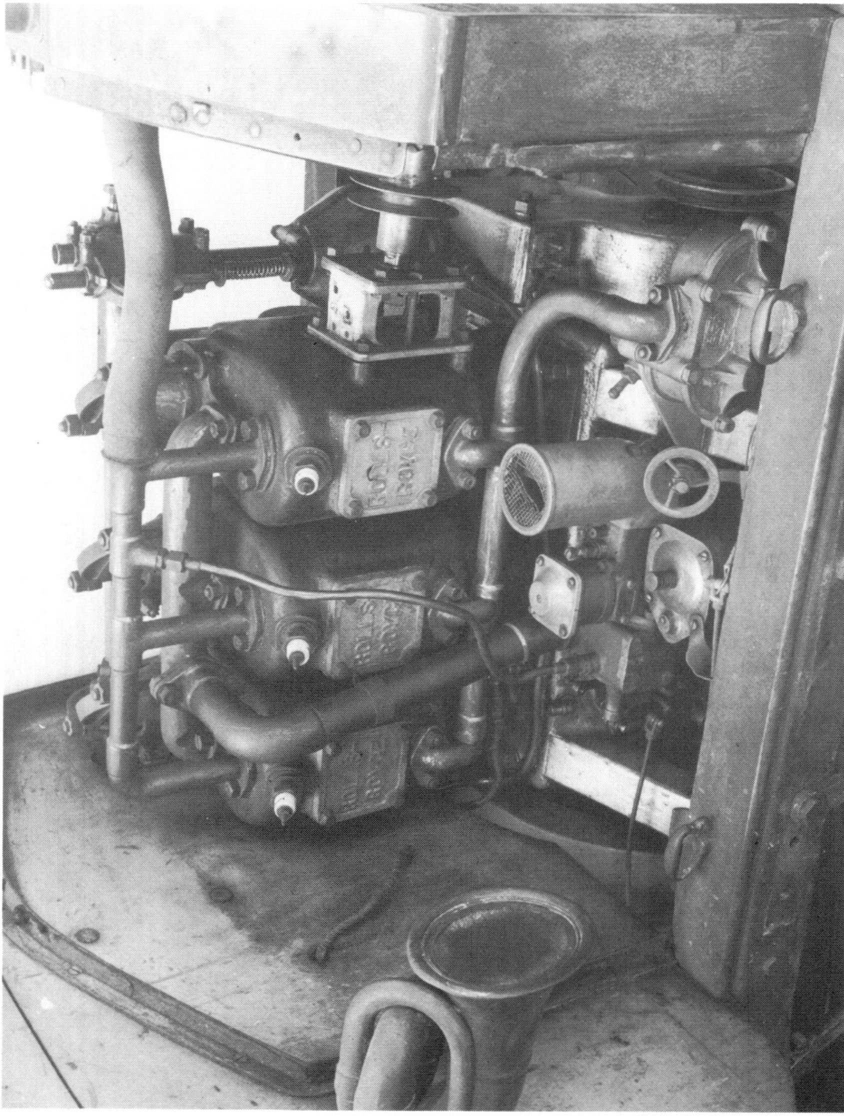

Figure 55 The offside view of the 15hp engine illustrates more clearly the cylinder top space problem referred to above.

Figure 56 The bolted-on balance weights on each web of the three cylinder crank look like those on the end webs of the Rolls-Royce twin, figure 52. This implies that there is effectively more balance weights per throw than on the twin, i.e. that a greater proportion of the reciprocating weight is balanced.

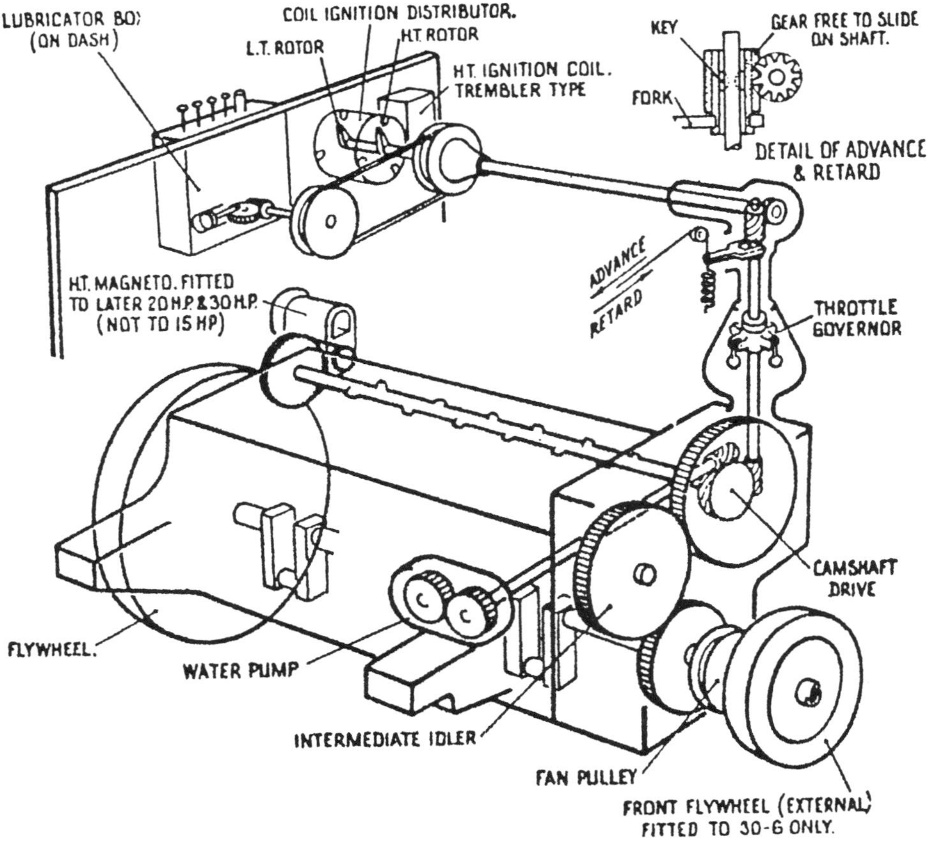

Figure 57 This is the auxiliary drive layout for the 15hp onwards. One skew gear at the front of the single camshaft drives the water pump via a cross shaft and the governor on a vertical shaft. A further pair of skew gears at the top of the vertical shaft drives an exposed shaft to the ignition distributor on the dashboard, and the oil pump via a belt. Engine ignition timing could be varied by endwise movement of the skew gear on splines on the vertical shaft. Control of this was by a very flimsy-looking exposed wire from the control on the dash.

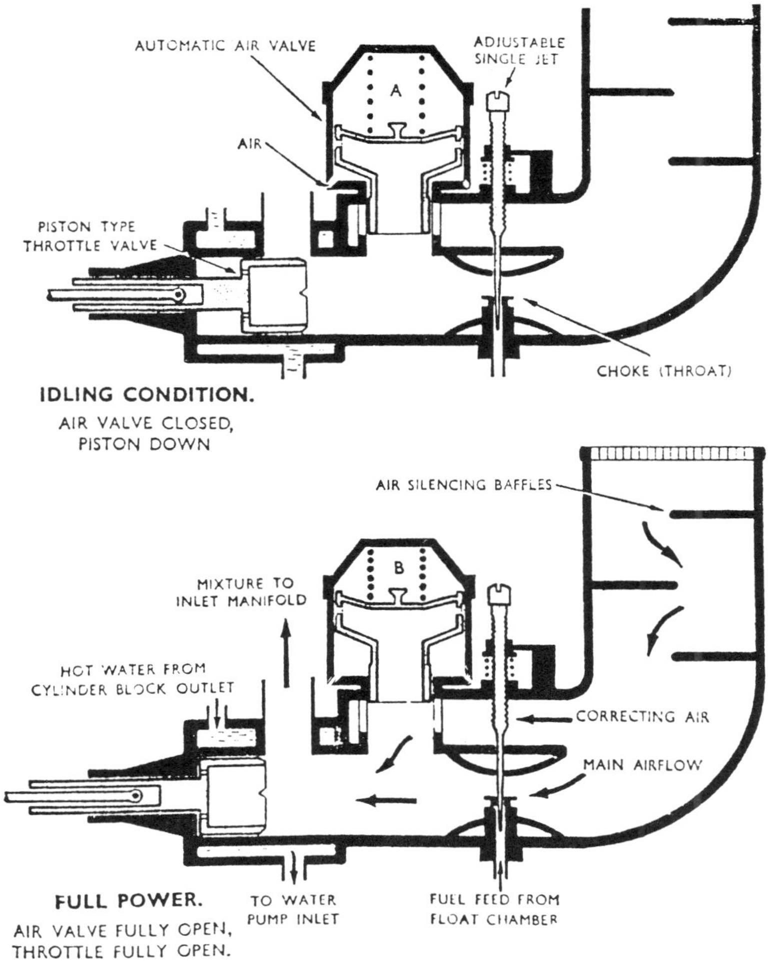

Figure 58 A petrol jet in a single choke tube gives a petrol-air mixture which gets richer with increasing air speed; early carburetters of this type were provided with a hand-operated extra air valve to allow the driver to compensate for this. This is irksome because ideally continuous variation of the extra air supply is necessary during acceleration, and if this is neglected there is a tendency to run over-rich rather than weak, to the detriment of the fuel consumption. R took an early decision that this state of affairs should be avoided by an arrangement which provided automatic compensation for mixture strength, improving both power and consumption figures and allowing the driver to concentrate on other matters. He achieved this by the air by-pass arrangement shown here, where the combination of spring and gravity forces on the air control piston can be arranged to provide any desired variation of mixture strength with speed. The deflector vanes in the air intake passage suggest an empirical solution to reducing air intake harmonics noise.

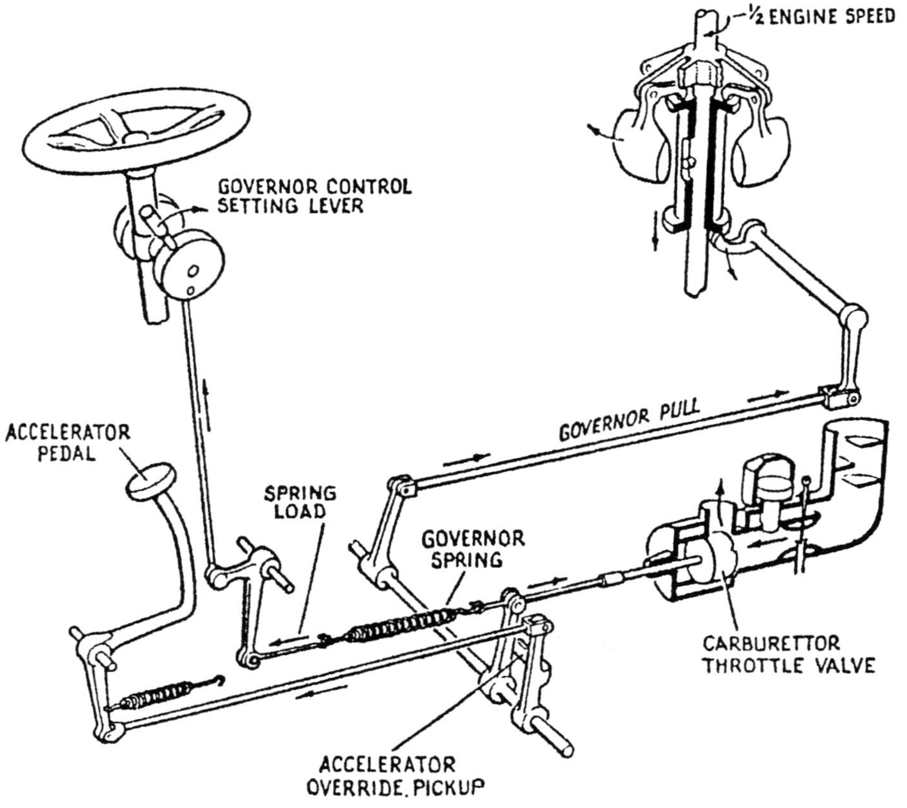

Figure 59 Automatic inlet valves and simple choke carburetters implied an optimum engine operating speed and this was recognised by the provision of a governor to control engine speed, leaving vehicle speed to be varied by gear changes. Some over-riding of this by the driver was necessary, even if only to try to achieve silent gear changes with the 'crash' type gearboxes. Mechanically operated inlet valves and automatic mixture strength control allowed satisfactory engine operation over a wide speed range and R adapted the governor to allow a hand lever to set a desired engine (and therefore road) speed, with the possibility of over-riding this by the accelerator pedal if extra speed was required. The resulting linkage is shown here.

Figure 60 The 30hp six cylinder engine. The bearing diameter that was adequate for the three and four cylinder engines gave too much torsional flexibility for the six.

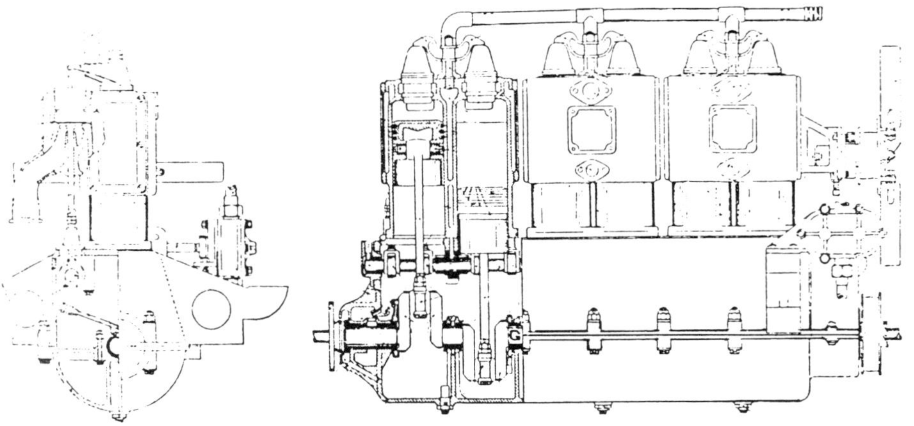

Figure 61 The sectional drawing of the 30hp engine confirms how much shorter the front end of the engine has become, by virtue of the auxiliary drive re-arrangement shown in (57). This drawing was obviously made after the 2 + 2 + 2 crank had been superseded by the now conventional 'mirror image about the middle' shaft with 120° crankpin separations.

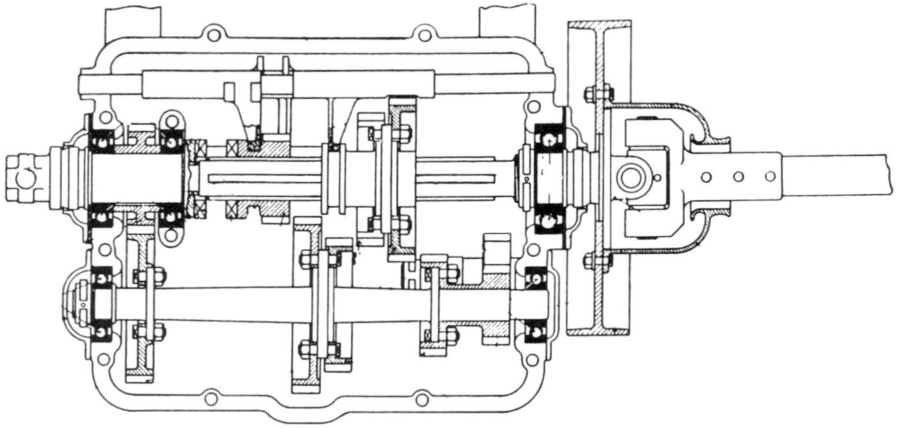

Figure 62 The provision of a four speed gearbox in no more space than the original three speeds meant using two selector shafts and two sliding units on the splined 3rd motion shaft. By using an overdrive top (R's "sprinting" gear) the arrangement of the original three speeds and reverse was retained, but the gate arrangement was unusual by today's standards in that both direct drive third and the second gears meant the selectors moved forward and therefore gear lever backwards.

Figure 63 A photograph of the casing and gears of the four speed gearbox for the T.T. cars. The original 10hp box had all plain bearings; the 15hp box retained plain bearings on the first motion shaft but went to ball bearings elsewhere; this box has ball bearings throughout.

Figure 64 The two cars filling up with their ration of fuel before the 1906 T.T. No. 4 is CSR's car.

Figure 65 CSR winning the 1906 T.T. race in the Isle of Man.

Figure 66 A photo of CSR and the winning car after the race.

Figure 67 This photograph of a six cylinder crankshaft being machined by Albert Wheatman, late in 1904 or early in 1905, shows that the original 2 + 2 + 2 crank arrangement was soon discarded in favour of the now universal arrangement of two three cylinder cranks back to back, which completely cancels out primary and secondary couples without the use of balance weights. In its final form the crank webs were not continued beyond the main bearings (61).

Figure 68 The chassis of the town carriage, designed to rival the briefly popular electric ones in silence and appearance and outdo them in performance.

Figure 69 The complete landaulet town carriage.

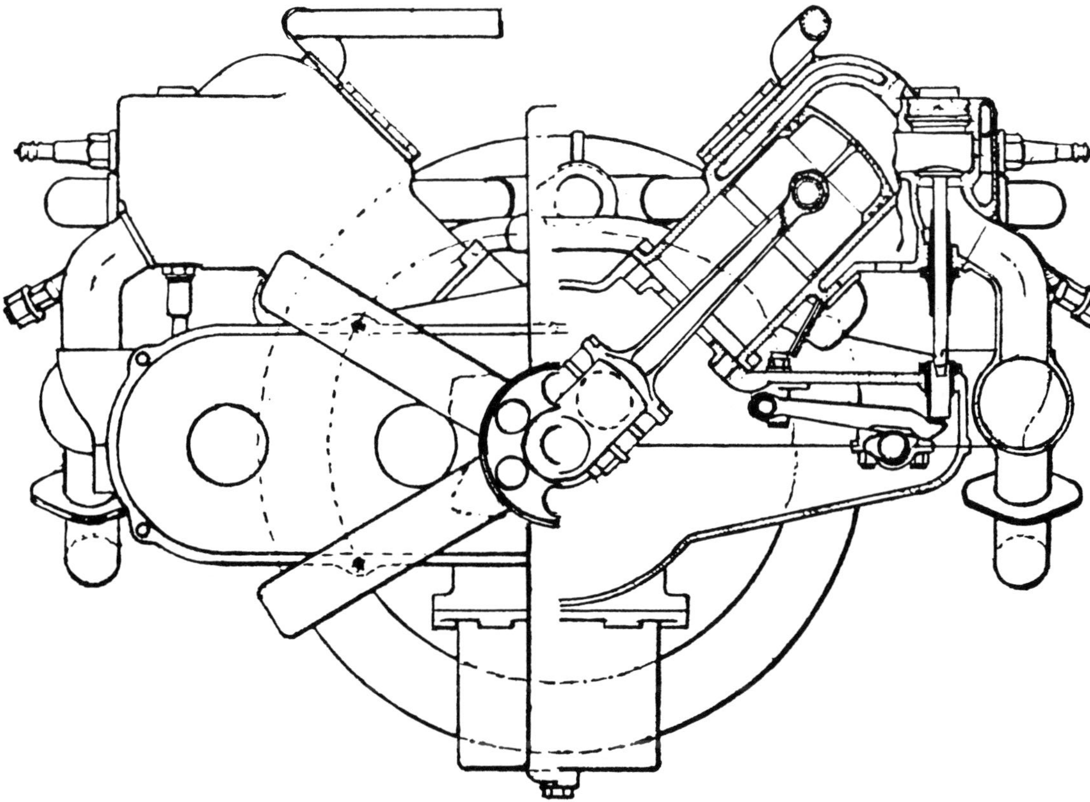

Figure 70 To achieve the appearance of the electric town carriage the engine had to be under the driver's feet. R made this possible by a short stroke engine of 3.25″ bore and stroke and therefore of 3.53 litre capacity. This front view shows how well R met the shallow overall depth requirement, some 22.25″, which compares with a flywheel diameter of 17.5″.

Figure 71 This 3/4 front view photograph is taken without the fan on the crankshaft nose. The auxiliary drive gearcase is prominent. Two major innovations for R were putting the main bearings into the upper half crankcase, and providing an engine oil pump and force feed to these bearings.

Figure 72 The plan view of the engine makes clear the continuing use of twin cylinder blocks, and the stagger between right and left hand blocks which results from the very early use of side by side connecting rods to achieve greater use of common parts and simplify those parts.

Figure 73 The short-lived town carriage engine was a stepping stone to this 40/50 six cylinder engine; force fed main bearings in the upper half crankcase, equal bore and stroke of 4.5″, and two blocks each of three cylinders to allow a reasonable length for the heavily loaded centre main bearing, are all features. The main bearings and crankpins too are now of sensible size and combine with the shorter stroke to avoid crankshaft torsional troubles. This is the "Autocar" sectional drawing.

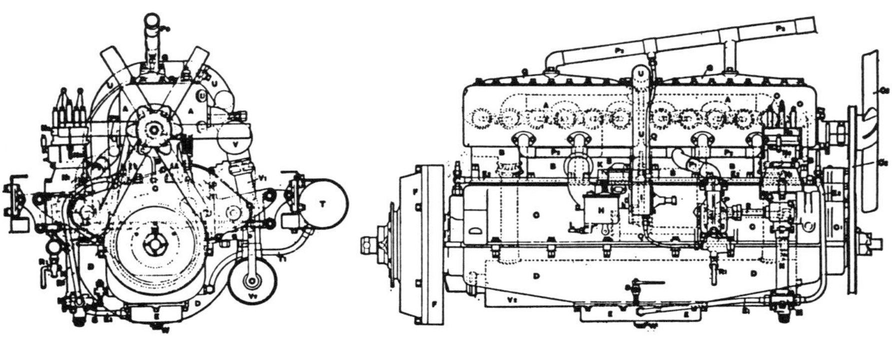

Figure 74 External, offside and front, views of this engine give more details. The improved Rolls-Royce carburetter on the offside is convenient for the hand controls and governor but gives a long inlet tract over the top of the engine. There are now two sparking plugs per cylinder and dual ignition.

Figure 75 The ability to carry large enclosed bodies and provide adequate performance was the basis of the 40/50 car design. Such an enclosed body is shown here.

Figure 76 The original 'Silver Ghost', so named by CJ and with nameplate below the windscreen, during its remarkable 15,000 mile observed trial.

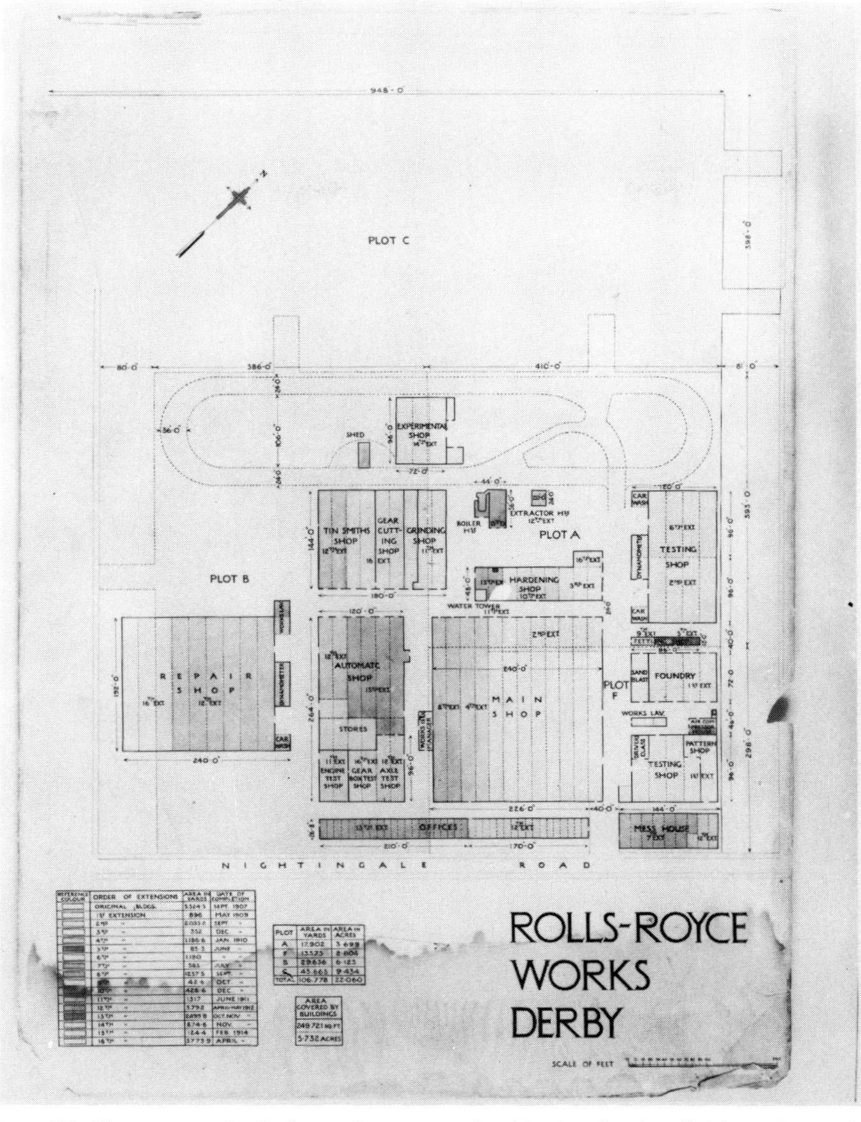

Figure 77 The move to the Derby works was completed in time for the official opening on 9th July 1908. An early plan of the works is shown here; the test track can be seen at the top.

Figure 78 This is an early view of the works from Nightingale Road.

Figure 79 The southwestern loop of the test track can be seen here.

Figure 80 The other, north-eastern, end of the test track is shown here.

Figure 81 In a later view of the works the office block is prominent.

Figure 82 The latest view of the works is an aerial view from the southeast; development on both sides of Nightingale Road has by now taken place.

Figure 83 One of the handsomer bodies on the 40/50 was the open tourer with the sloping bonnet, seen here.

Figure 84 The 40/50 chassis was continually developed during its long lifetime; it would have looked like this when the Spanish Grand Prix races were run. Obvious features in this examples are the cantilever rear spring and the addition of the double kingpost frame stiffeners. Not so obvious are the changes from a transmission brake behind the gearbox to a second set of shoes in the drums at the back wheels, now of double diameter to accommodate the two sets, one within the other; and the torque tube on the rear axle instead of the torque arm.

Figure 85 The suggestion of running two 40/50s in the 1913 Spanish Grand Prix came from Count Carlos de Salamanca. In this photograph he is on the left; next to him is CJ's wife, then CJ himself, holding the dog. R himself is unmistakeable; the final figure on the right is Lord Northcliffe.

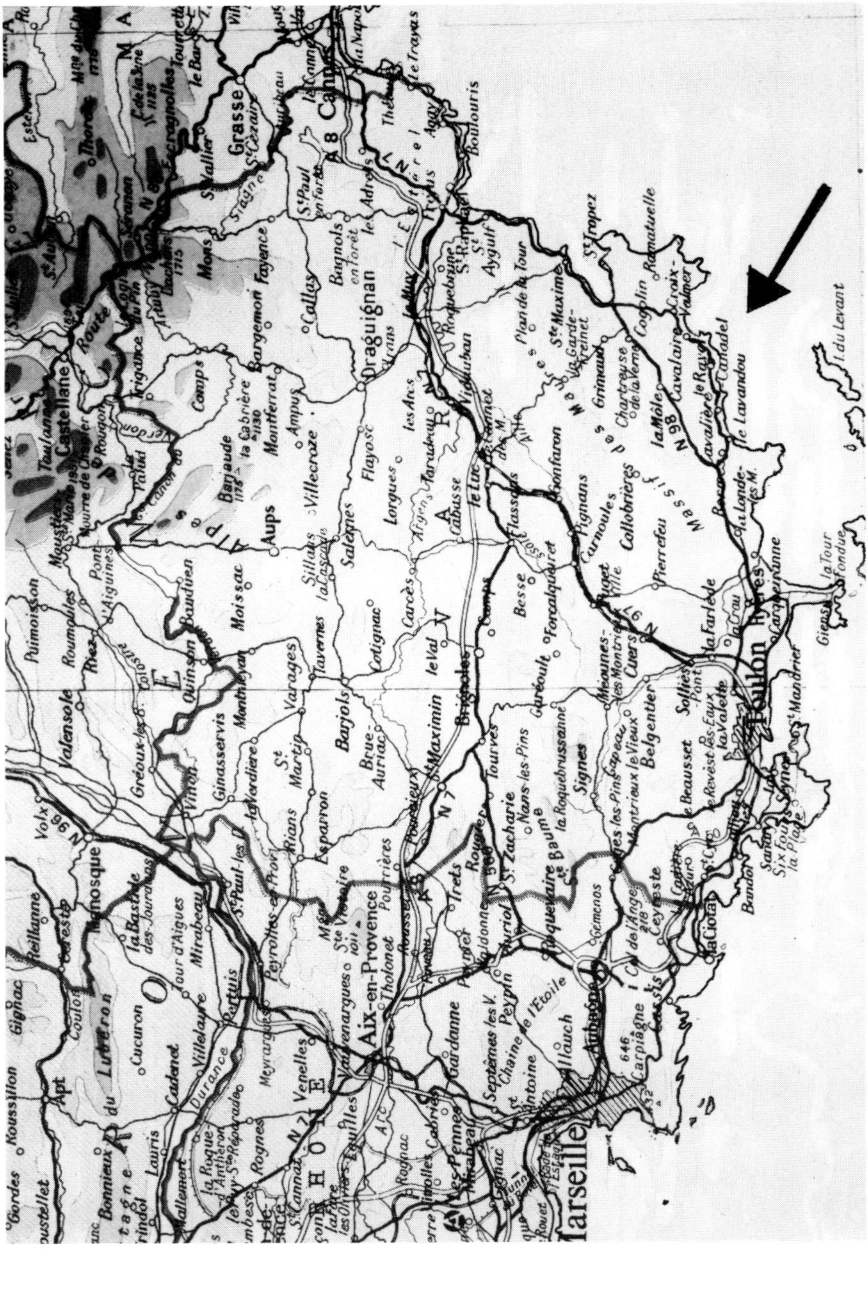

Figure 86 This map shows the geographical location of Le Canadel (Canadel-sur-Mer), towards the western end of the French Riviera.

Figure 87 A recent picture of CJ's Villa Jaune.

Figure 88 A prospect of Le Canadel: the large building on the right is the hotel; then there are, from upper right to lower left, CJ's Villa Jaune, R's Villa Mimosa and Le Bureau.

Figure 89 Don Eyre's drawing of Le Bureau.

Figure 90 R's painting of his Villa Mimosa.

Figure 91 Royce's garden and home, Elmstead, West Wittering.

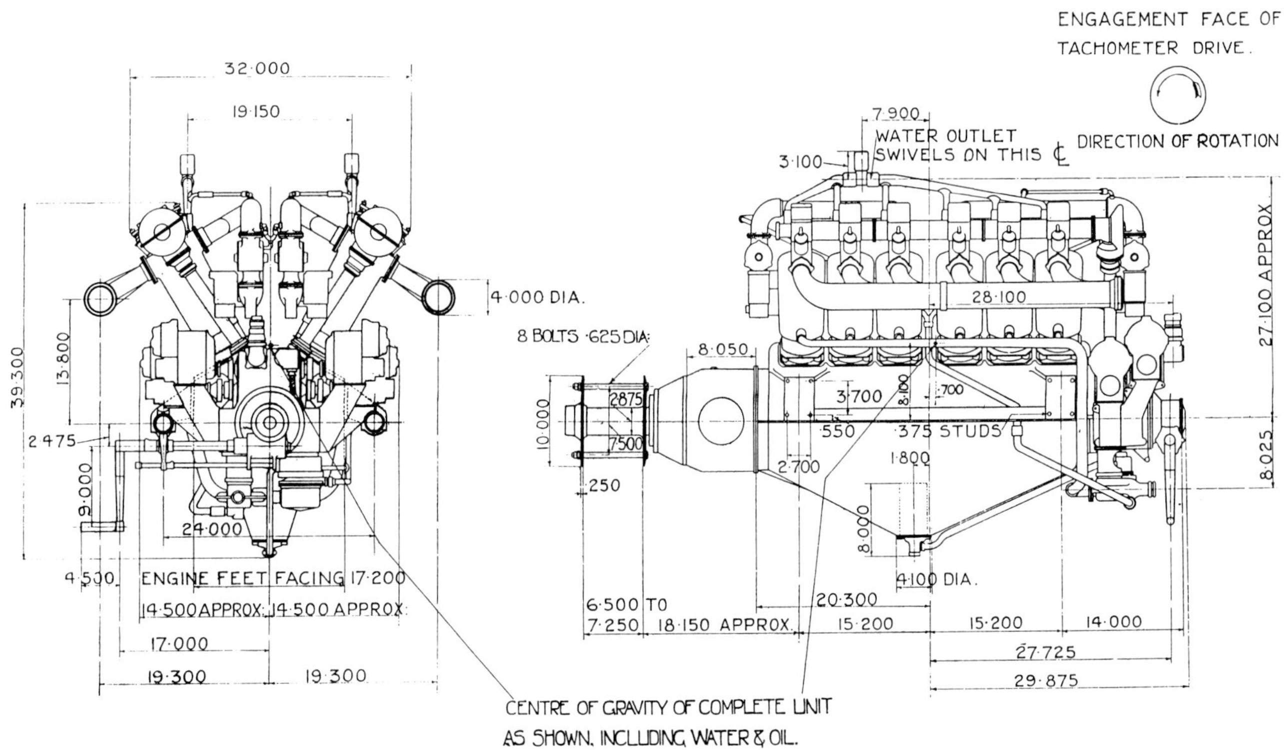

Figure 92 This drawing is of the Eagle VIII, effectively the final version of this engine, operational before the end of World War I.

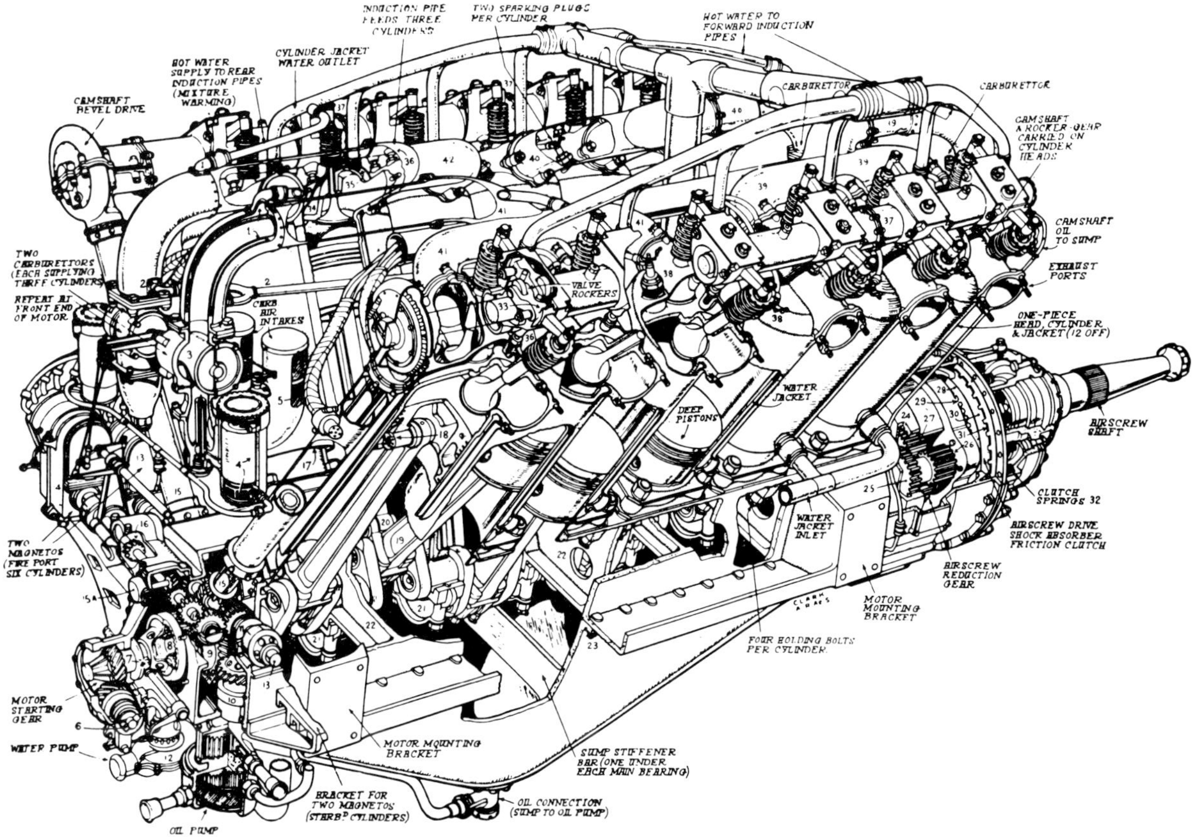

Figure 93 This is a sectional perspective drawing of the Eagle VIII.

Figure 94 A photo of the Eagle VIII from the propeller end.

Figure 95 An Eagle VIII engine on test; two 40/50 radiators are used to cool it.

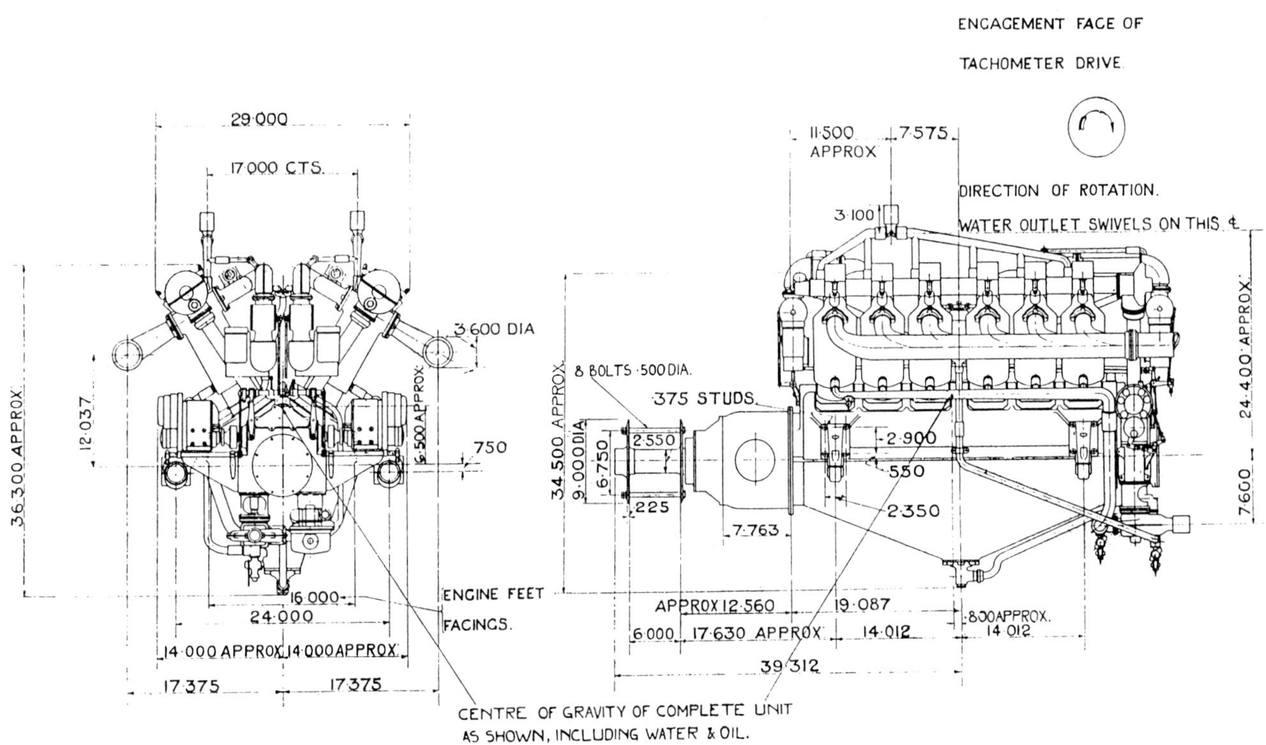

Figure 96 A drawing of the Falcon III, the scaled down version of the Eagle used in the Bristol fighter.

Figure 97 A photo of the Falcon III.

Figure 98 A photo of the installation of the Hawk engine in a 'Blimp', from the propeller end. The Hawk was effectively a single bank of the Falcon engine, but with a longer stroke. The combination of less power, resulting smaller diameter propeller, and longer stroke implying slower rotational speed, meant that a reduction gear was not needed on this engine.

Figure 99 The nacelle used for the Hawk installation. In both pictures the exhaust system can be clearly seen, an early example of exhaust tuning; the evenly spaced exhaust pulses from the two halves of the engine are kept separate until the best merging place is reached. The final fishtail does its best to convert a bark into a hiss.

Figure 100 The Condor engine after its redesign. It was fundamentally a scaled-up Eagle, but with four valves per cylinder. Disposing these radially on a part spherical combustion chamber required a valve gear which was the subject of patents. The angle between the valves was limited by the need for those on adjacent cylinders to clear each other. The first example of cross bolts in the main bearing caps will be noted. Lack of rigidity limited its further development.

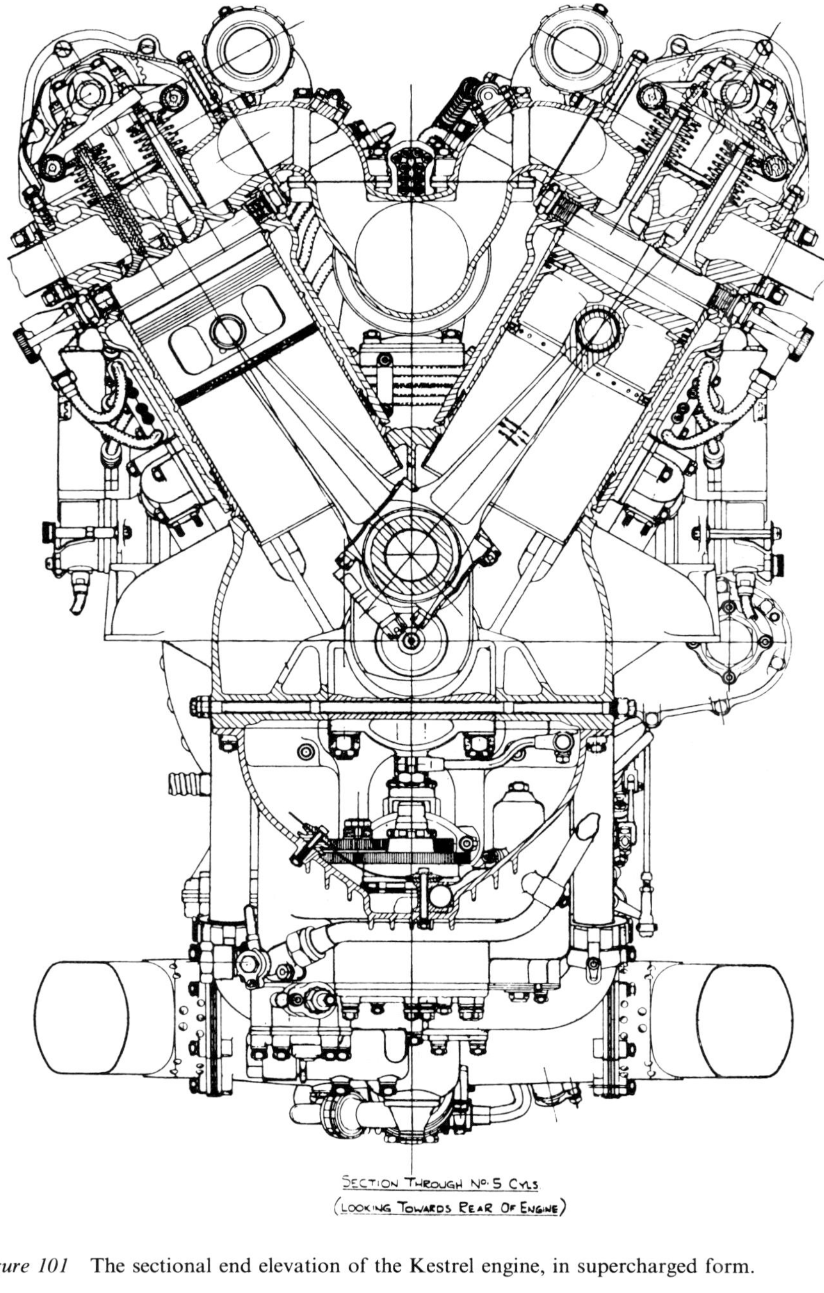

Figure 101 The sectional end elevation of the Kestrel engine, in supercharged form.

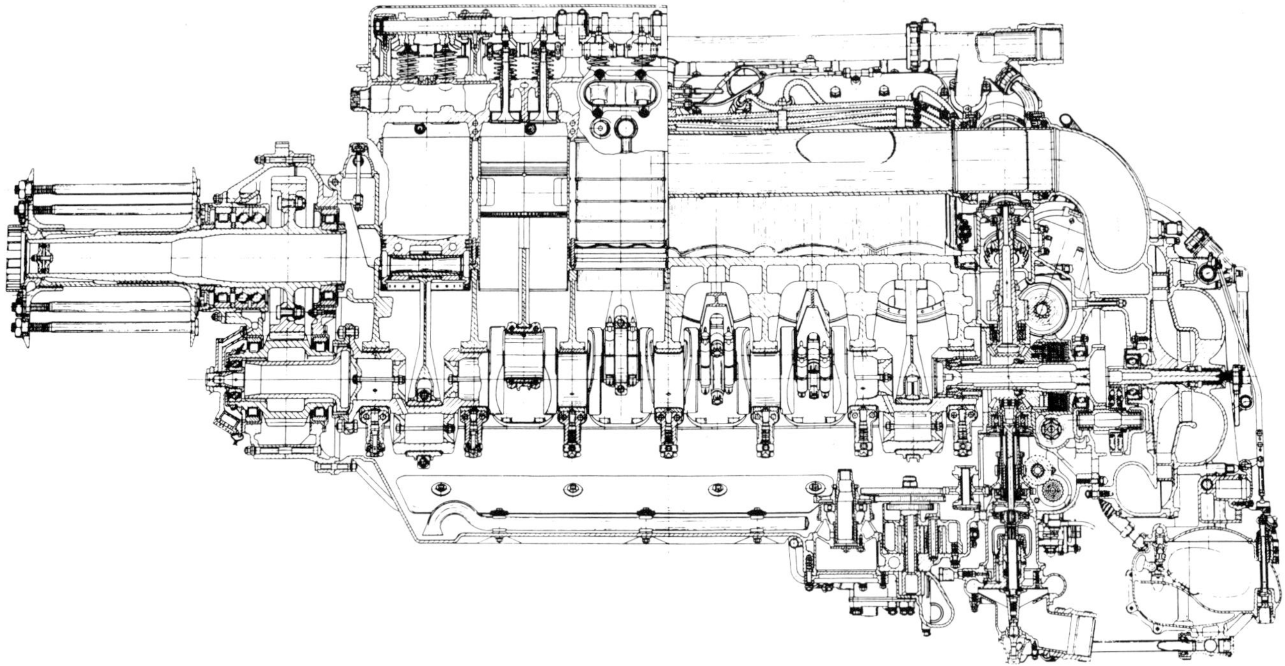

Figure 102 A sectional side elevation of the same engine. The fork and blade connecting rods will be noticed, and the continuous block for each bank of six cylinders. Four valves per cylinder are retained, but they are parallel and in a flat head. Although this is the first production Rolls-Royce aero engine with cylinder blocks rather than individual cylinders, each of the four banks of four cylinders of the Eagle XVI were complete blocks.

Figure 103 The Kestrel had been the first Rolls-Royce aero engine to be supercharged as standard. Its final development was the Peregrine, shown here, which had a downdraught carburetter, increased boost to take advantage of the availability of 87 octane fuel, and some Merlin features. It saw some service in the Westland Whirlwind.

Figure 104 The Vulture shown here was fundamentally four of the Kestrel six cylinder blocks mounted on a single crankcase with a common crankshaft. Although theoretically in production the big end construction proved to be unreliable and the engine was withdrawn from service. The Avro Manchester with two Vultures was modified to and superseded by the Lancaster with four Merlins.

Figure 105 The Buzzard, shown here, was a 1.2 to 1 scale up from the Kestrel. It was known originally as the H engine, was made in fairly limited numbers but is famous as the basis from which the two versions of the R engine, for the 1929 and 1931 Schneider Trophy races, were developed.

Figure 106 This is the R engine; the most obvious difference between this and the Buzzard is the very large supercharger.

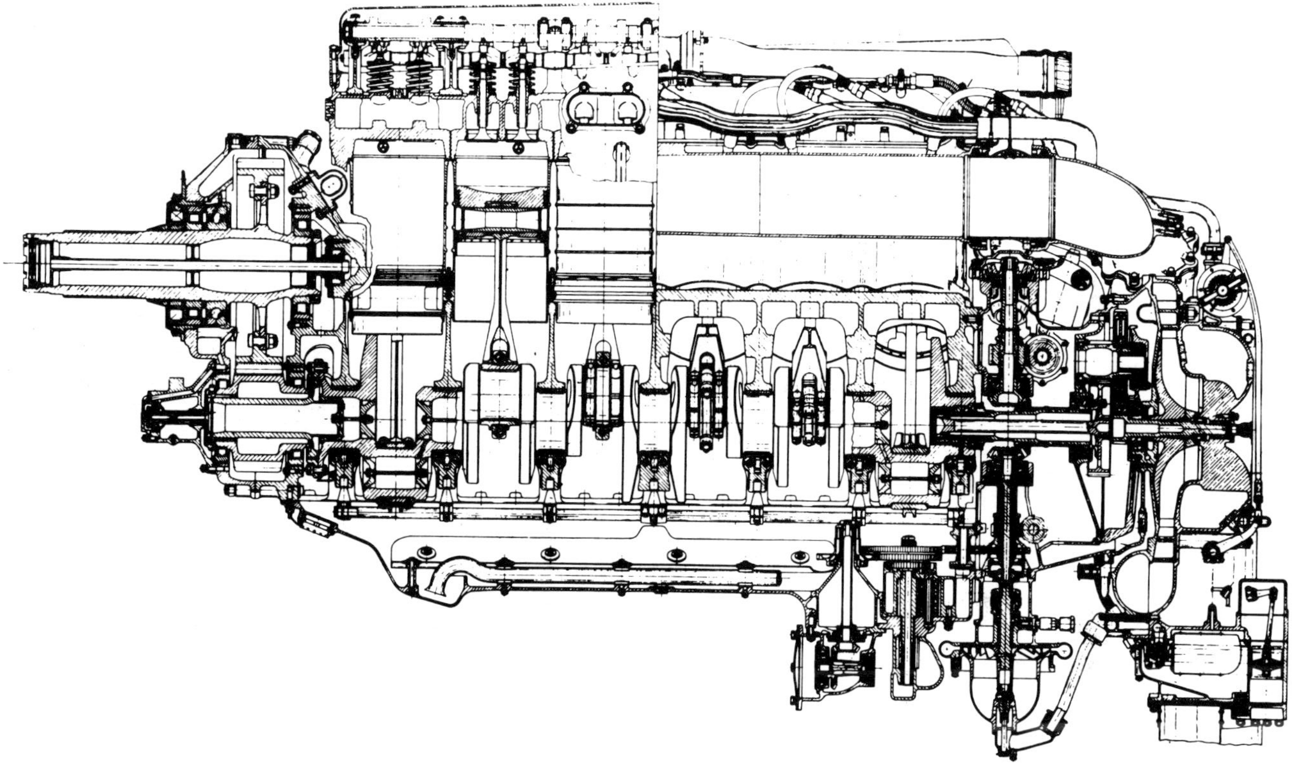

Figure 107 The last aero engine whose design was started in R's lifetime was the Merlin, of which this is a sectional elevation, of the Series II.

Figure 108 This is a photo of the Series III of the Merlin. Although single cylinder development of the Merlin started with the ramp head, which had two of the four valves per cylinder parallel to the cylinder axis, the other two at an angle, this proved to have weaknesses and the design fairly quickly reverted to the Kestrel combustion chamber shape. The Merlin was made in very large quantities in World War II: at the Rolls-Royce factory in Derby, in shadow factories under Rolls-Royce control, by Ford at Manchester and by Packard in America.

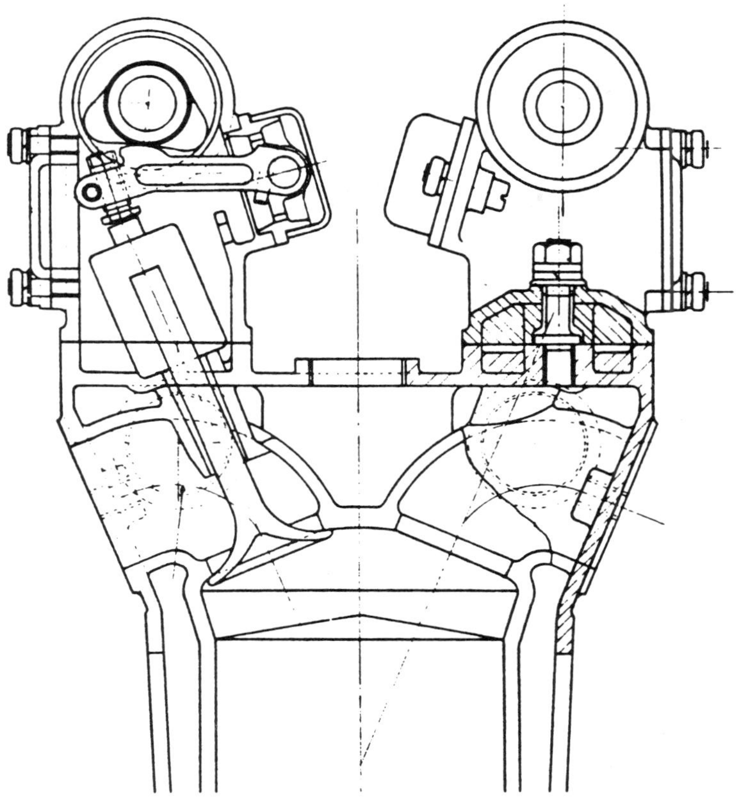

Cylinder head 'Goshawk I'

Figure 109 As this cross section shows, the first experimental post-war 20hp car engine had valves operated by twin overhead camshafts.

Figure 110 The auxiliary drive gears, with the spring drive-cum-slipper flywheel detached for clarity, of the production 20hp engine.

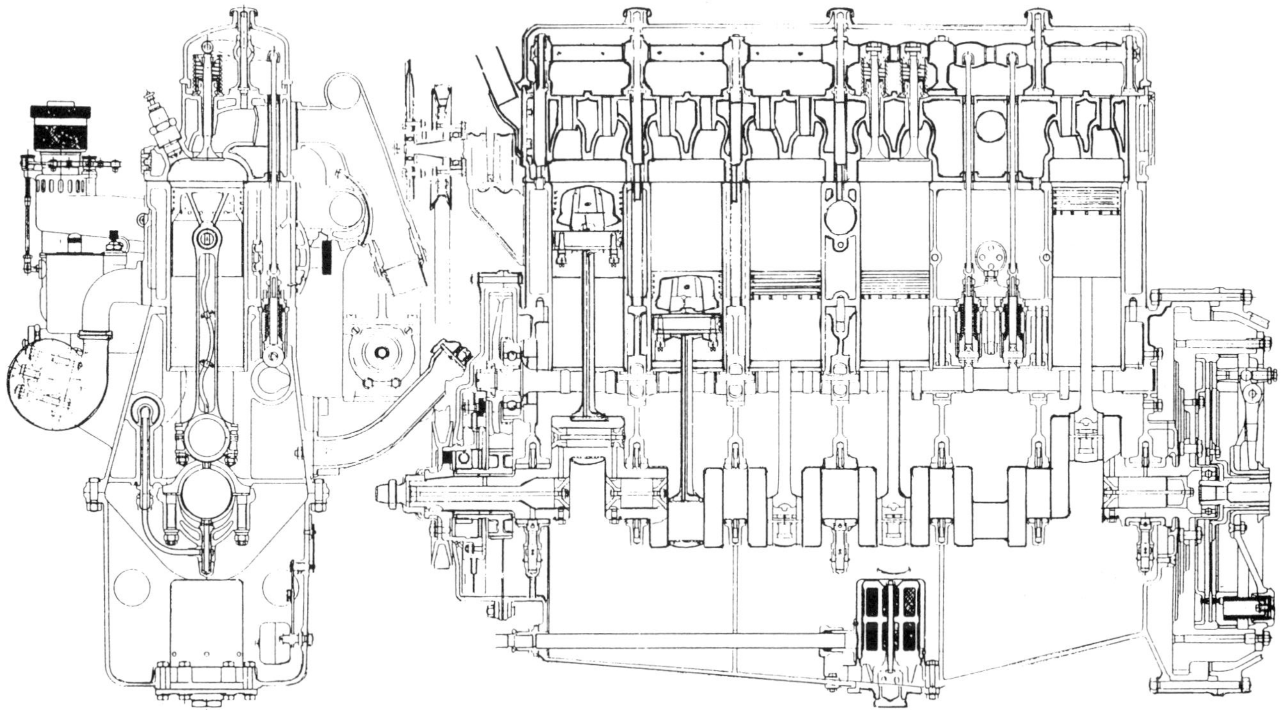

Figure 111 Sectional side and end elevations of the 20hp engine with its final cylinder head arrangement.

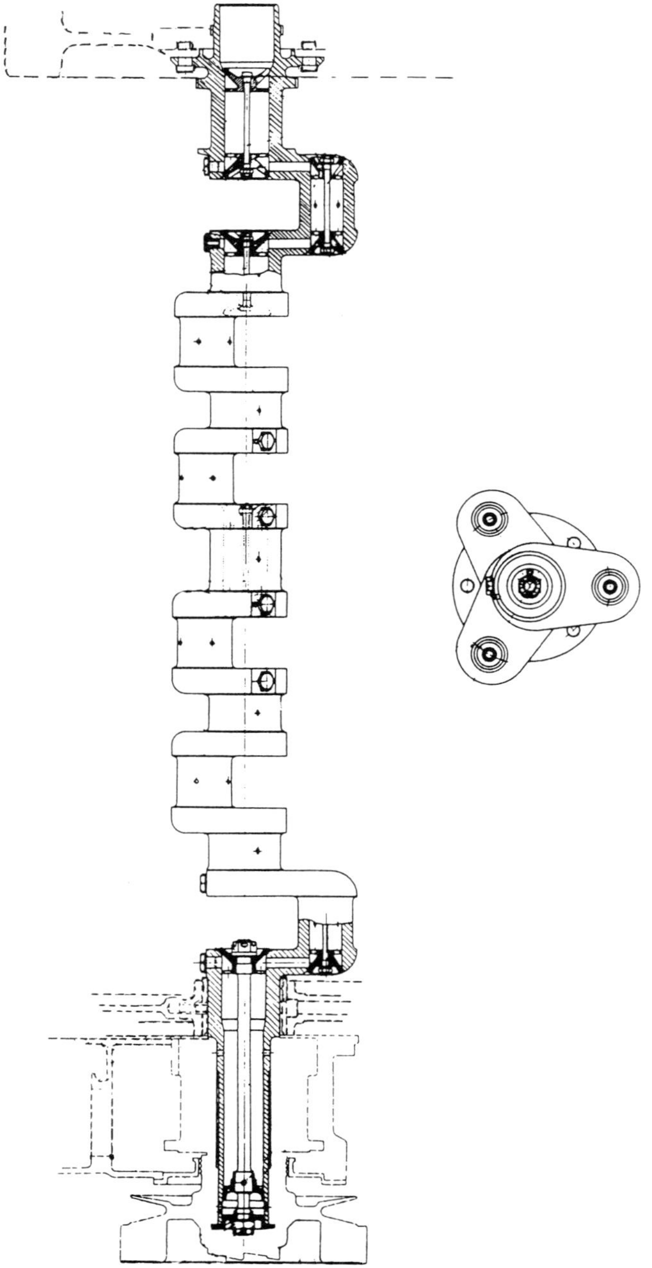

Figure 112 The 20hp crankshaft.

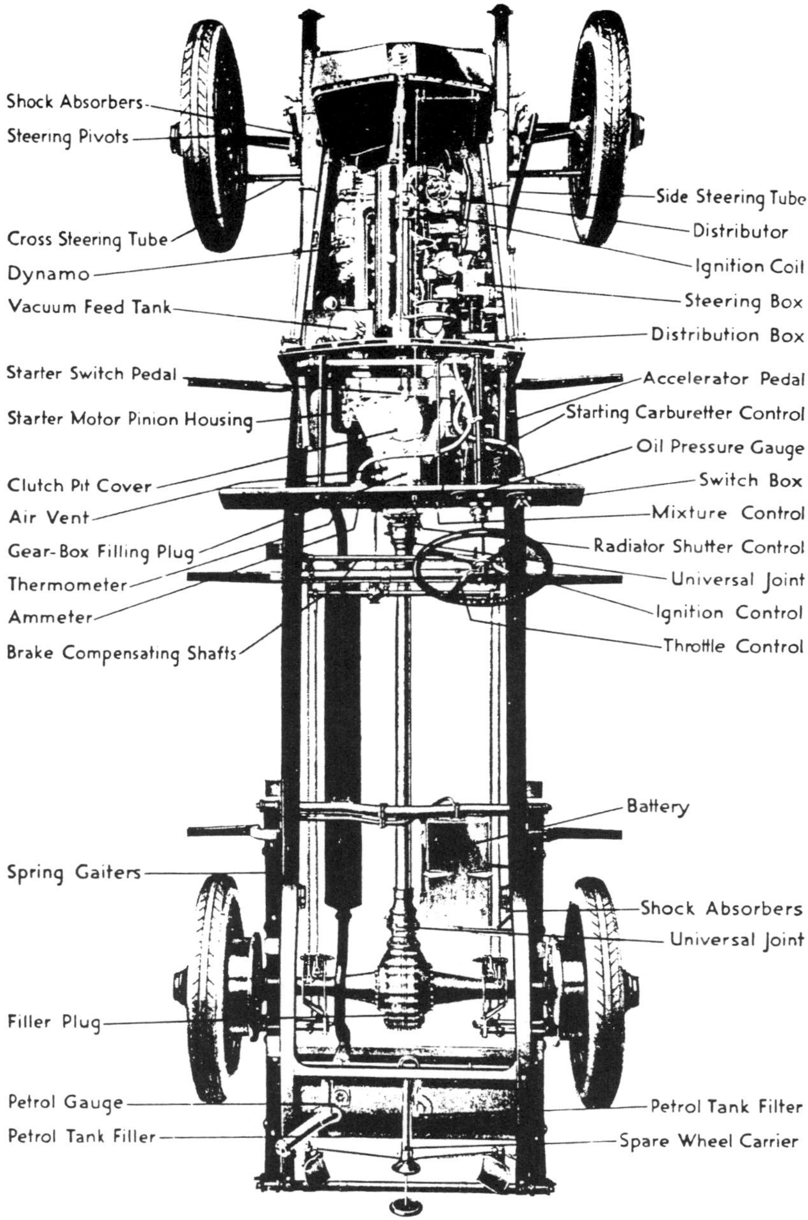

Figure 113 A plan view of the original 20hp chassis, 3 speed gearbox, centre change, no front brakes.

Figure 114 A photo of the 20hp chassis.

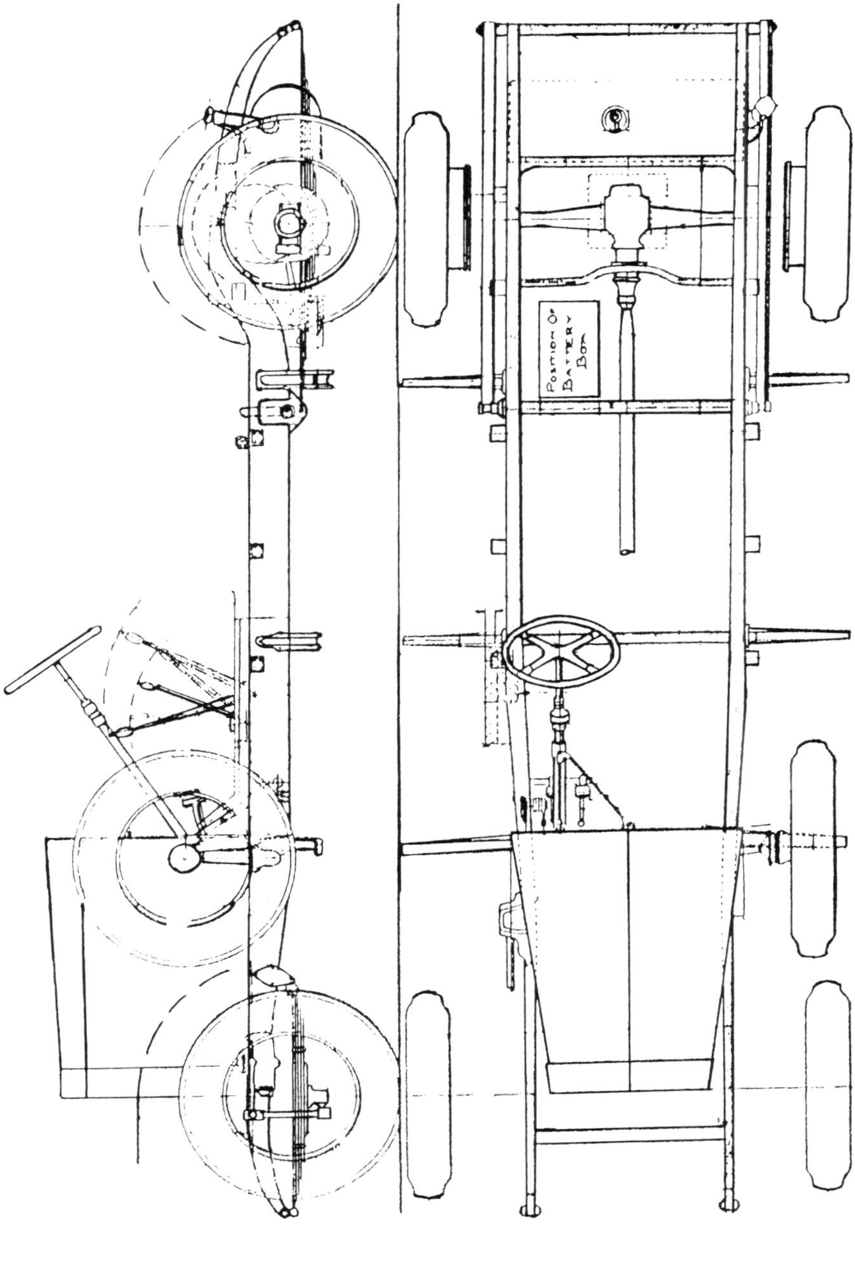

Figure 115 The coachbuilders drawing of the 20hp car.

Figure 116 An open four seater body on an early chassis; this was the type that would come nearest to R's original stipulation of no more than 35cwt fully laden. It would appeal to the owner-driver.

Figure 117 Front wheel brakes were fitted to the 40/50 from 1924. An operating shaft each side, in front of the axle beam and carried in bearings from it, had a lever at its inner end pulled by a rod from the servo. A lever at the outer end of this shaft pulled down a lever on the brake camshaft via a rod. The effects of wheel steering were minimised by having the top end of the operating rod as close to the kingpin centreline as possible. Later axles incorporated the operating shaft bearings.

Figure 118 This is the mechanical servo, to assist the driver put on the four wheel brakes; in this instance it is seen as applied to a left-hand drive car. The pedal rod A pulls on the end of lever A_1; the connection between levers A_1 and A_2 is by sloping dog faces, which produce an axial force to engage the servo friction faces inside O, the servo body, which is rotated from the back wheels via the gearbox output shaft and skew gears. Rod A_3 transmits the pedal force to a lever on a cross-shaft, on its way to operate the rear brakes. Rods R and F are connected to the servo front plate, pulled round by the friction faces, and have appropriate lost motion provision so that whichever way the car is going the equaliser lever T is pulled upwards. The upper end of the actual equaliser bar, whose pivot can move forward or backward by swinging of its support T, to allow force equalisation, adds to the pedal rod effort to operate the rear brakes via rod T_r; the lower end operates the front brakes via rod T_f. The angling of the rods F and R will slightly affect the equalisation, giving rather more rear braking when going backwards.

Figure 119 The engine fitted to the "New Phantom" is the first overhead valve 40/50; it still has two blocks of three cylinders, united by the single cylinder head. This has inlet and exhaust systems on the same side, in this respect resembling the new 20hp engine.

Figure 120 The town carriage de luxe of the early thirties - a Phantom II.

Figure 121 This Thrupp & Maberly body on a 20hp chassis illustrates the type of enclosed body which was likely to make a nonsense of R's overall weight limit, laden, of 35cwt for this car.

Figure 122 This is the 1929 version of the engine - gearbox unit of the 20/25.

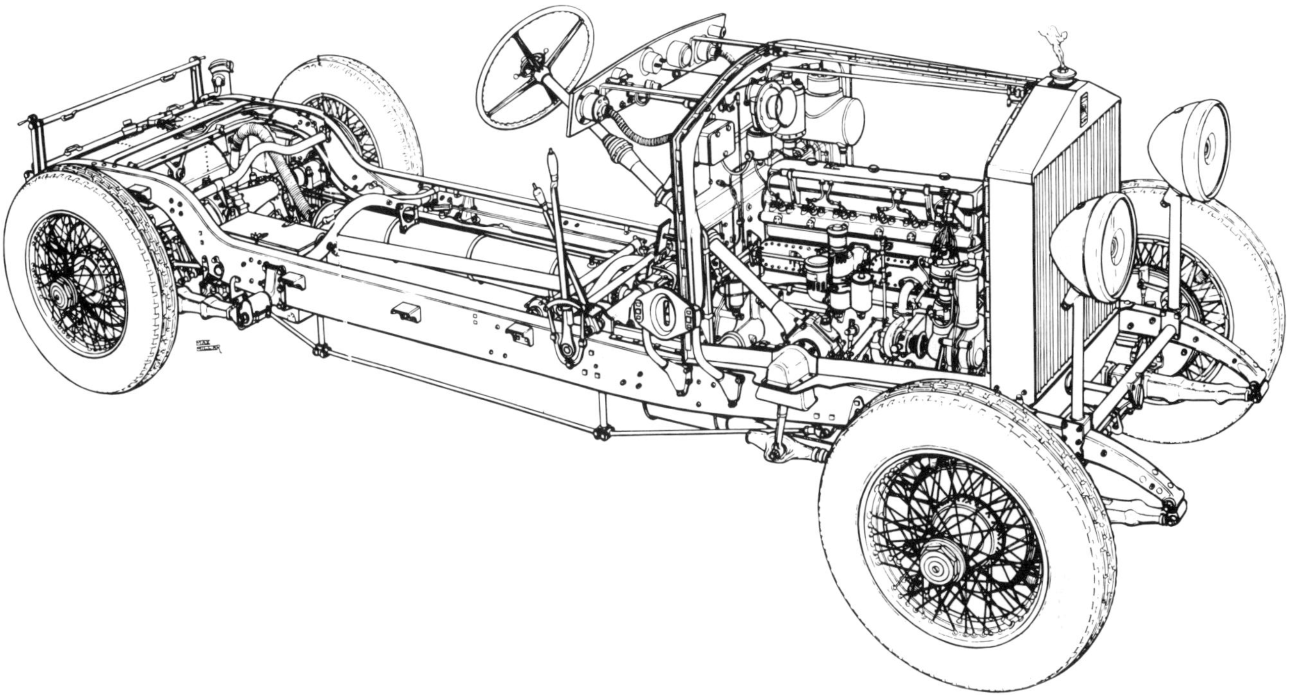

Figure 123 The 1929 model 20/25. It has front wheel brakes, with servo, four speed gearbox with side change, and the double kingpost frame sidemember stiffeners.

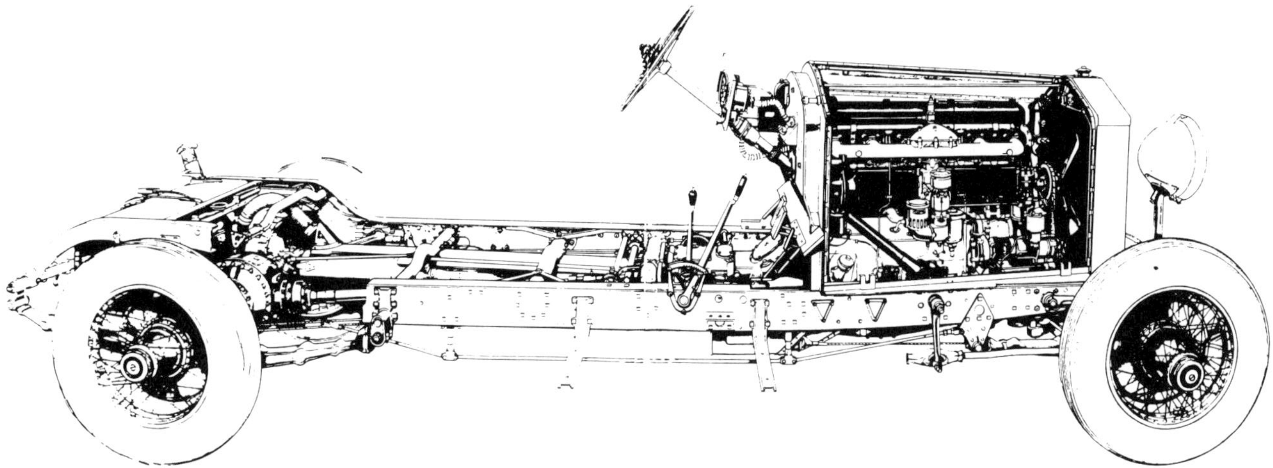

Figure 124 The Phantom II chassis has many differences from the previous 40/50s: engine and gearbox in one unit, open propeller shaft, semi-elliptic springs all round, and others less obvious.

Figure 125 The Phantom II Continental saloon on the shorter wheelbase was the attempted answer to the performance of the 8 litre Bentley. The car is seen outside Villa Mimosa at Le Canadel with Henry Royce and Sir James Percy.

Figure 126 The Phanton II engine was the final version of the large six cylinder overhead valve engine, with cross flow cylinder head.

Figure 127 The arrangement for taking front axle brake torque relied upon a parallelogram linkage: the rear part of the spring, anchored at the back and therefore shackled at the front, provided one link, the lever of the hydraulic damper the other. The triangular frame Q was pivoted to the axle outside the spring, to allow chassis roll and single wheel bump. The brake operating cable can be seen in the foreground; beyond and above the frame it is possible to see how closely the end of the brake camshaft lever comes to being in line with the kingpin.

Figure 128 An example of the 3.5 litre (Derby) Bentley introduced in 1933.

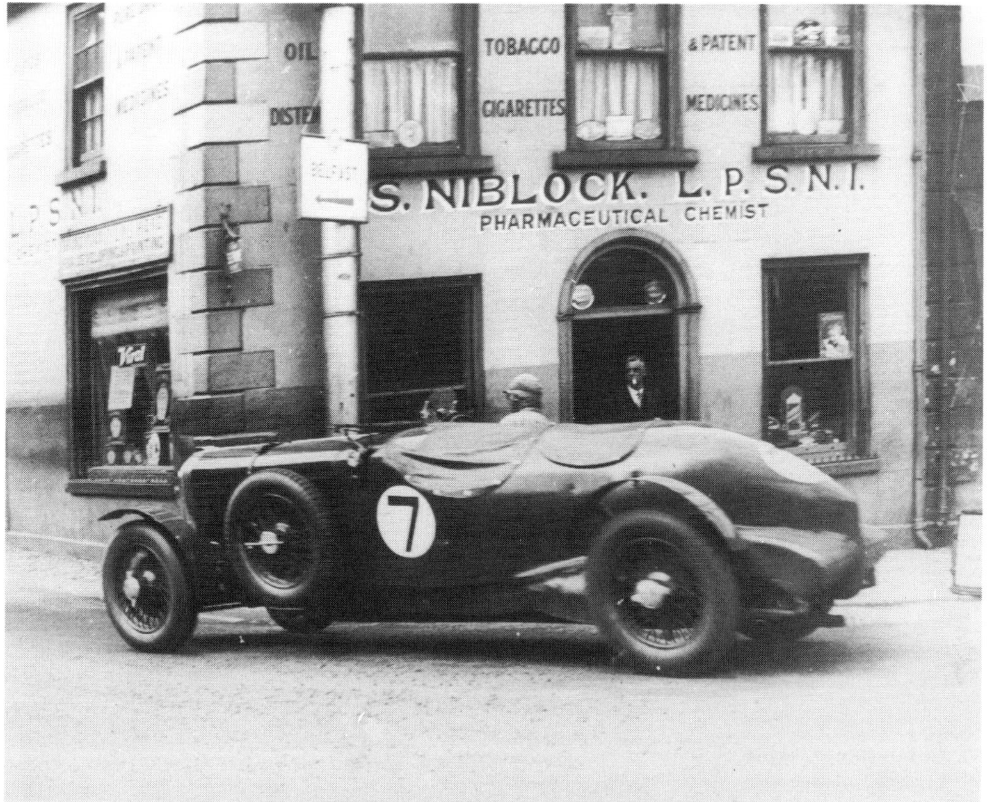

Figure 129 One of the factory prepared E.R.Hall (Derby) cars which put up notable performances in successive Ulster T.T. races. The Rolls-Royce Enthusiasts Club Bulletin No.143, March-April 1984, tells the story of these cars.

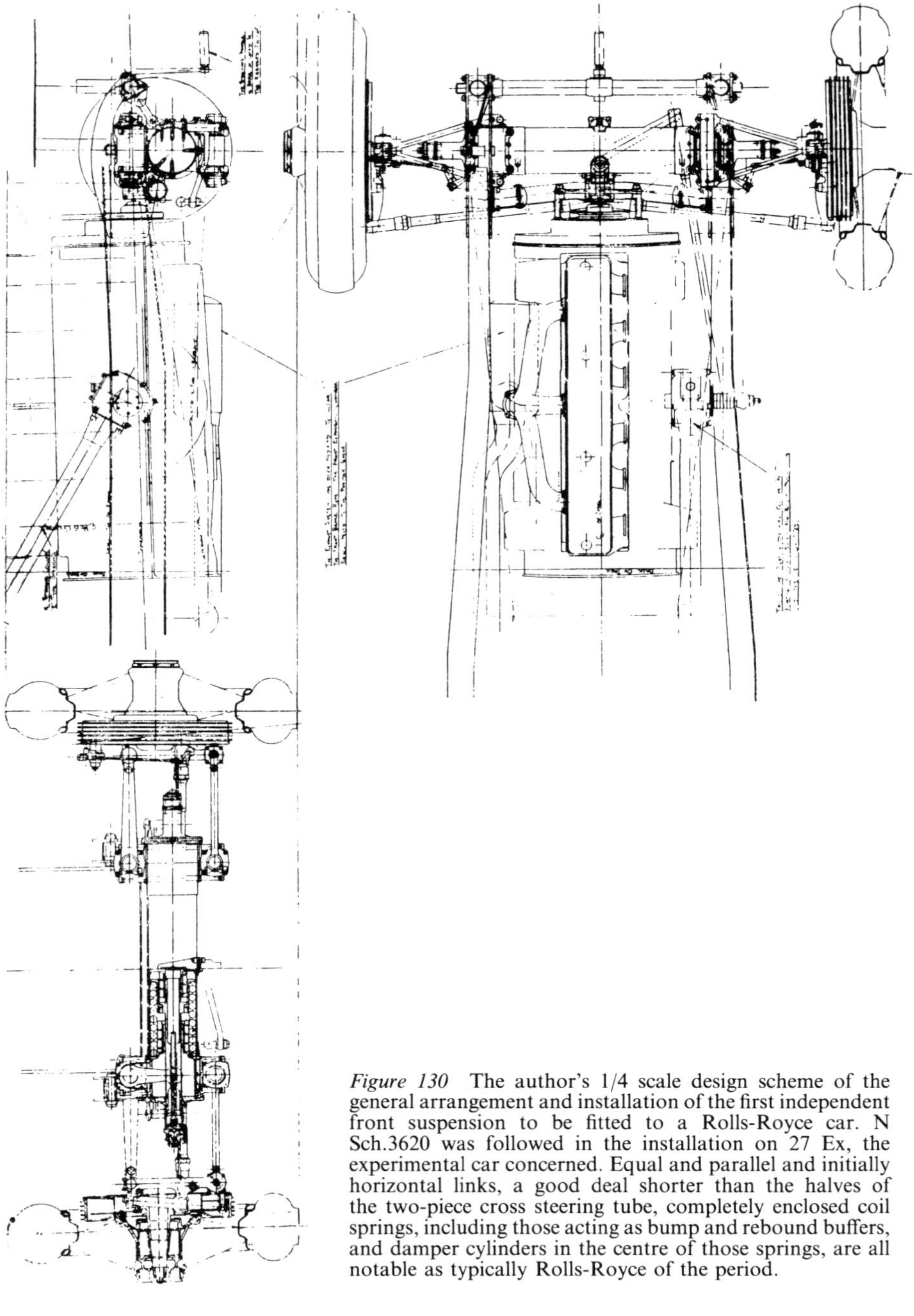

Figure 130 The author's 1/4 scale design scheme of the general arrangement and installation of the first independent front suspension to be fitted to a Rolls-Royce car. N Sch.3620 was followed in the installation on 27 Ex, the experimental car concerned. Equal and parallel and initially horizontal links, a good deal shorter than the halves of the two-piece cross steering tube, completely enclosed coil springs, including those acting as bump and rebound buffers, and damper cylinders in the centre of those springs, are all notable as typically Rolls-Royce of the period.

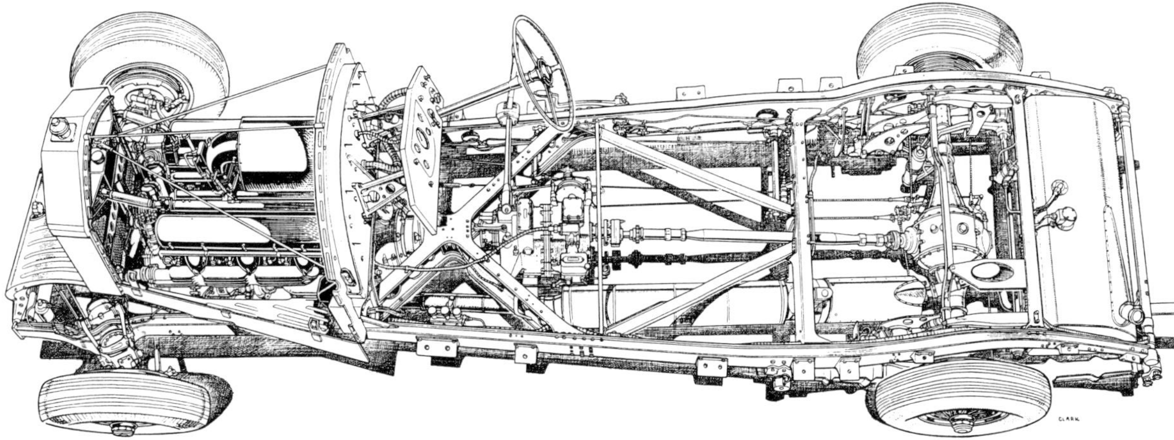

Figure 131 The Phantom III frame; there is a short frame cross, well forward, for torsional stiffness at the front end; the independent front suspension is fully built in, as later figures will show, and the kinks near the front are to accommodate this. The frame was not liked by the works, presenting problems in maintaining accuracy.

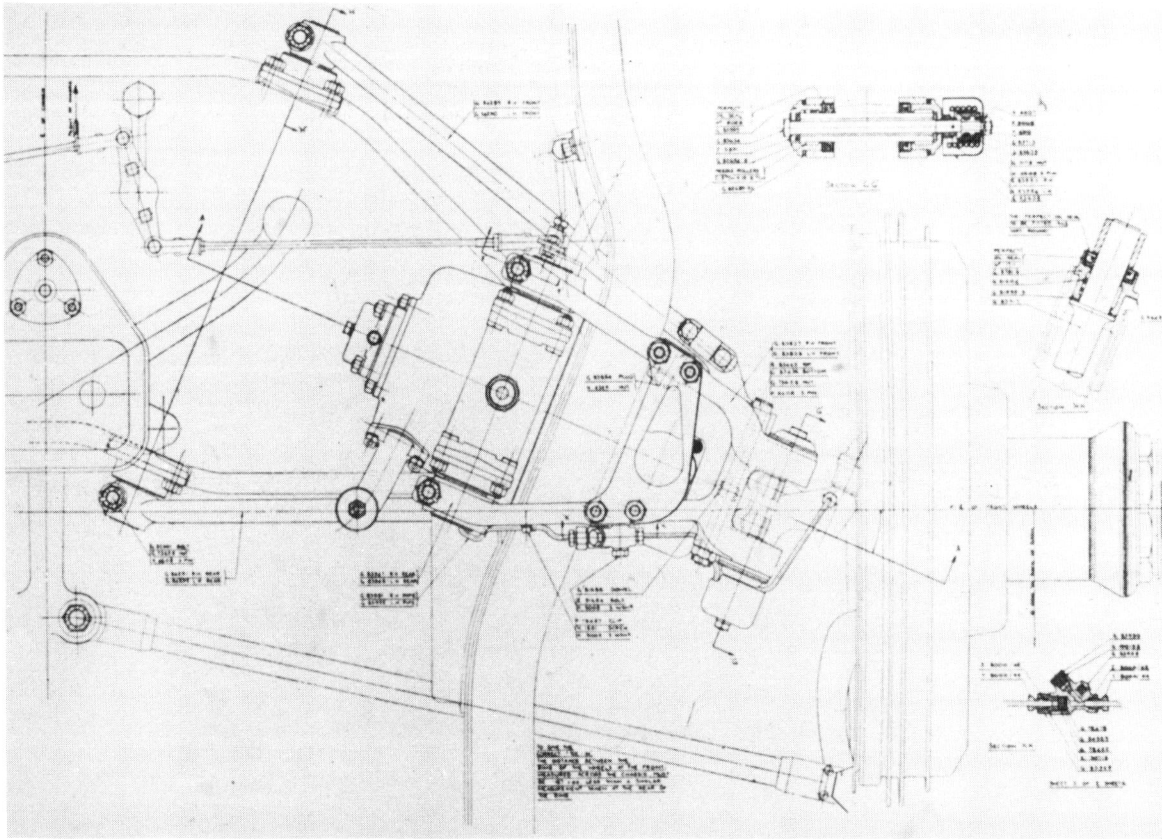

Figure 132 This is the plan view of the Phantom III suspension. The longer lower triangle levers now provide a better match for the separate cross steering tube lengths.

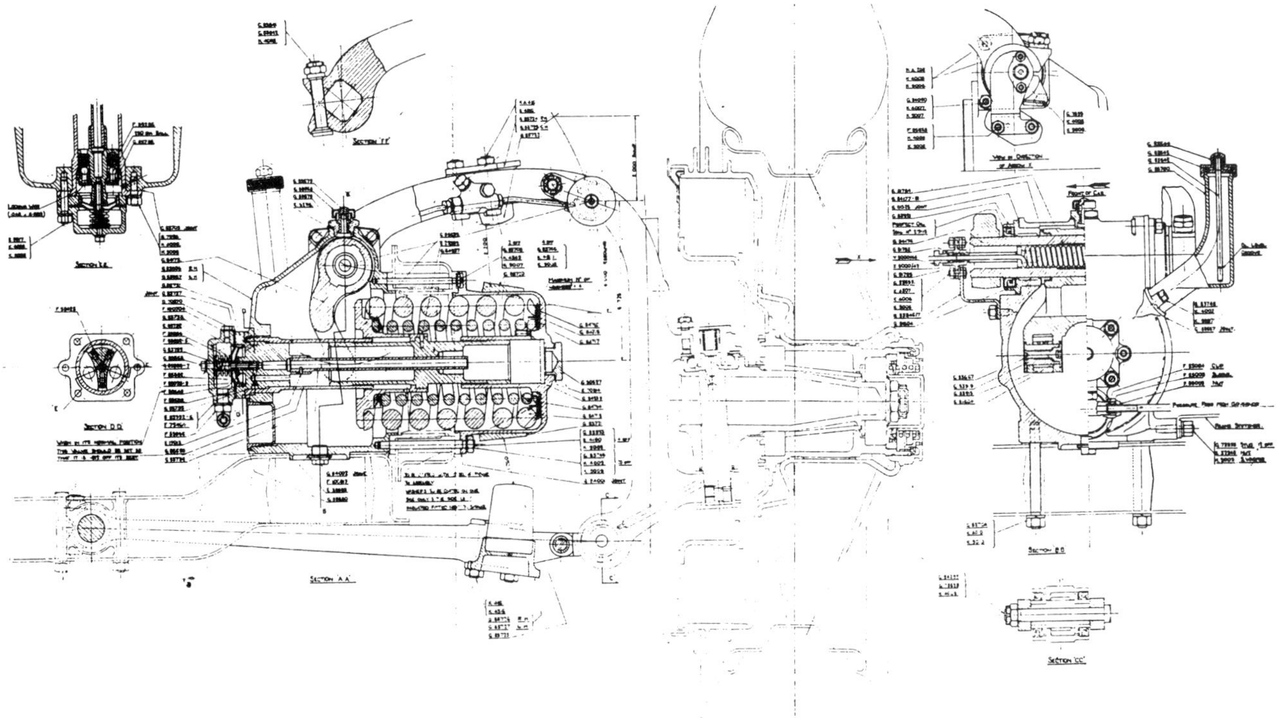

Figure 133 A view along the axis of the suspension links. The diverging ends of the upper and lower links (A to C series) provide a good deal of swing axle effect and were the result of last-minute changes; D series cars had a revised linkage which avoided this effect. External rubber bump and rebound buffers are now incorporated, all the spring space being required for the double suspension springs when the much greater suspension movements are allowed for. The damper is still inside the springs and incorporates the control unit.

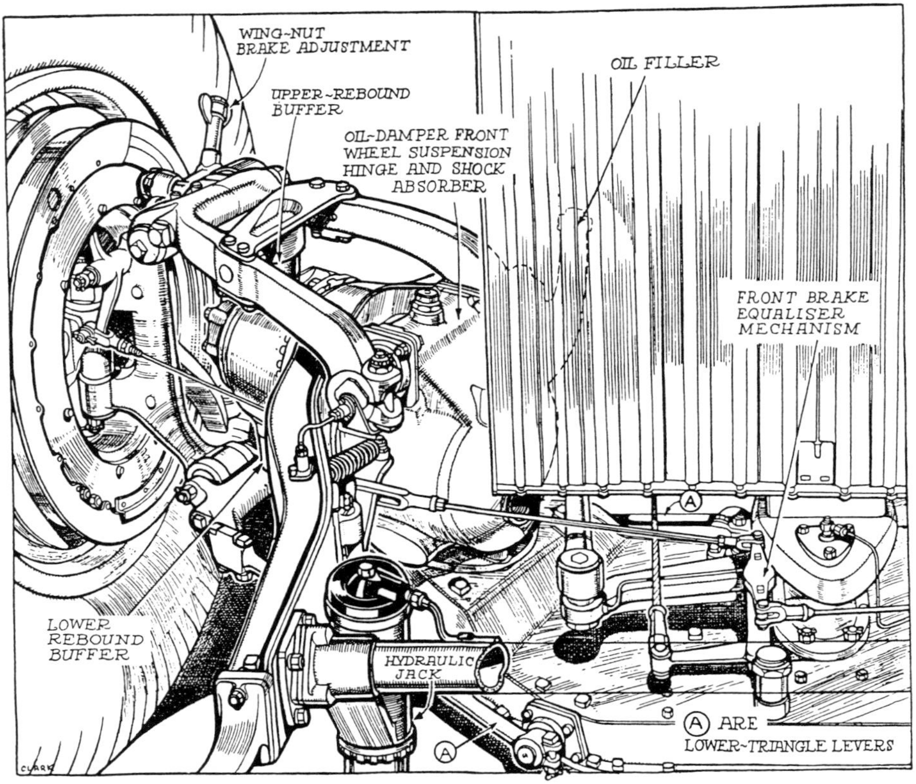

Figure 134 A perspective view of the Phantom III independent front suspension from the front makes some of its features clearer; the front end of the frame side member is omitted for clarity.

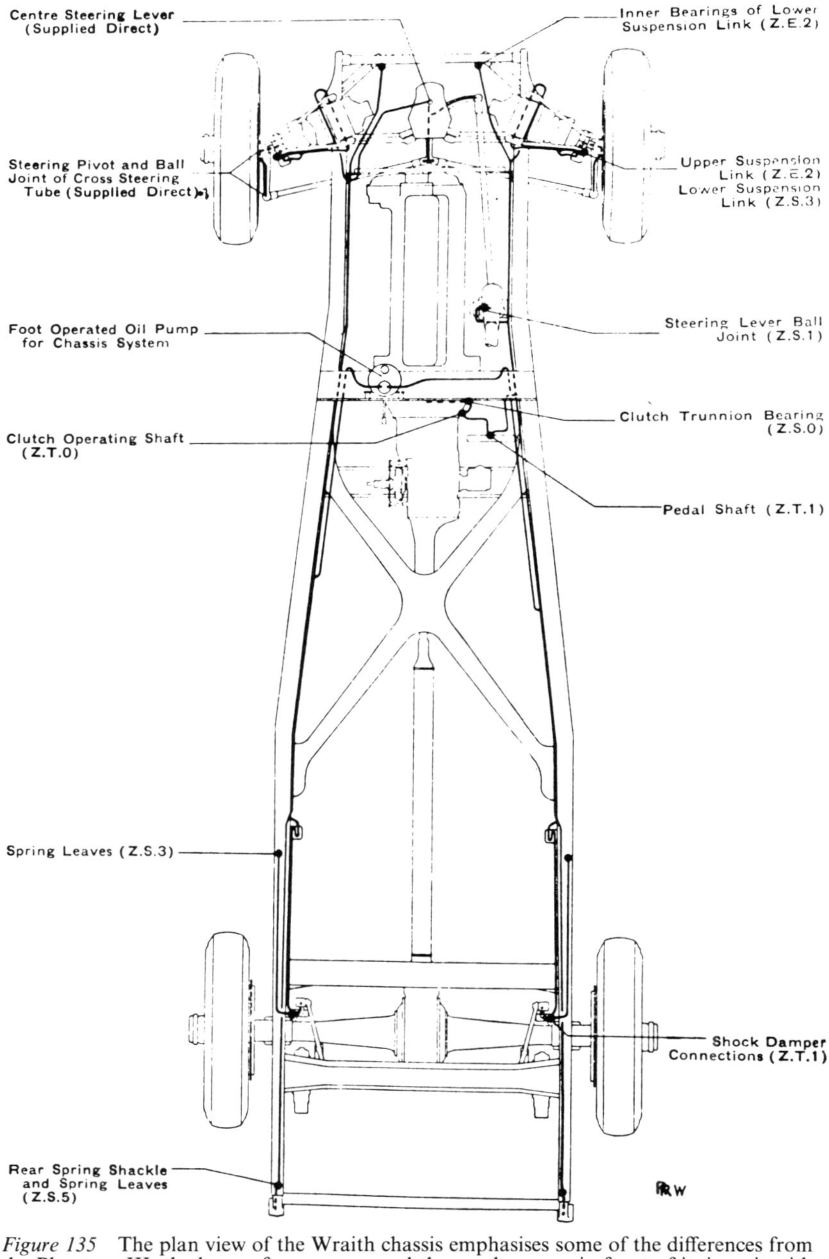

Figure 135 The plan view of the Wraith chassis emphasises some of the differences from the Phantom III: the longer frame cross, and the gearbox now in front of it, in unit with the engine, are two of the more obvious ones.

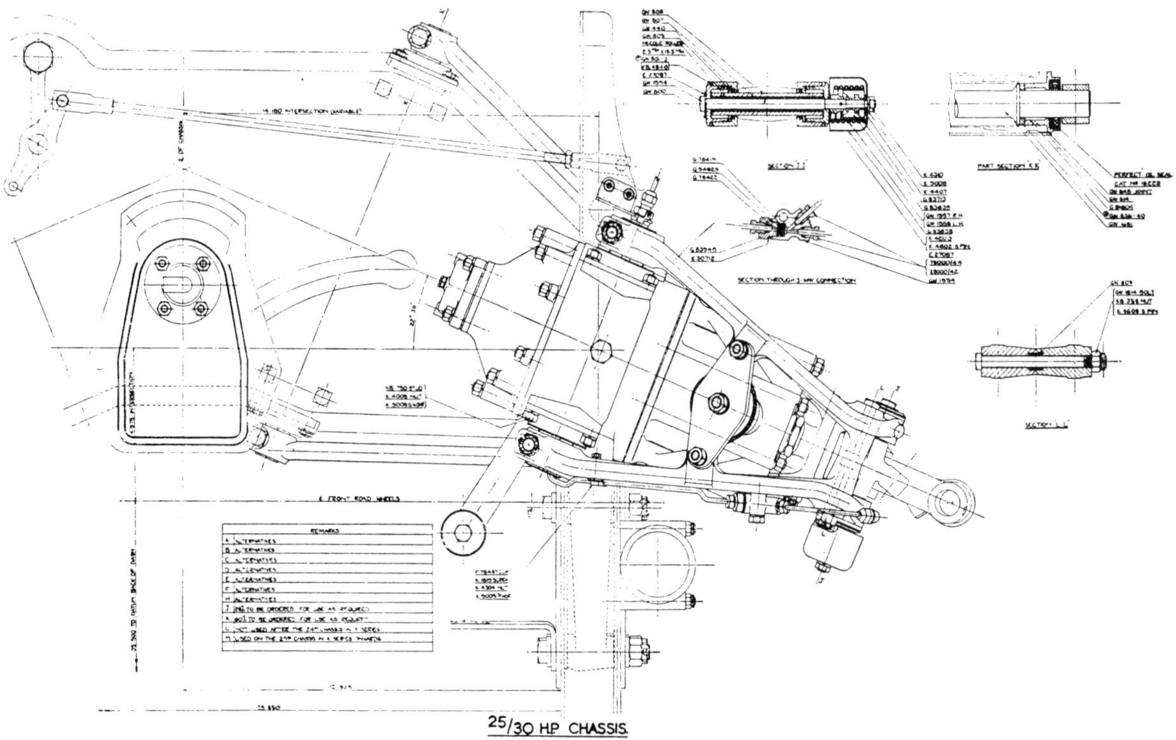

Figure 136 The plan view of the Wraith independent front suspension shows similarities to and differences from its predecessor. The frame members themselves stop short at about the front wheel centre line and the triangle levers are now symmetrical in plan and are therefore not handed. A steel forging still provides the main body but inner cover and spring housing are aluminium castings.

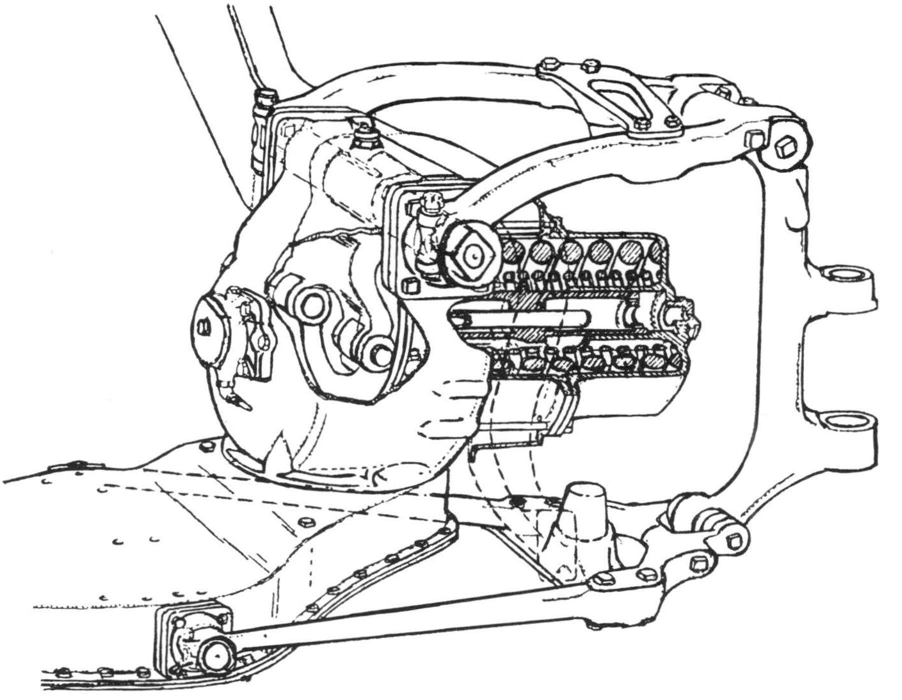

Figure 137 In this view we can see that the main body of the suspension has a tail enclosed within the front of the boxed side member and secured to it by four bolts. Frame manufacture was considerably simplified by this feature, and the suspension could be made up as a complete unit.

Figure 138 The view of the Wraith front wheel, on lock, shows the very handy brake adjustment common to all Derby cars at that time; the wing nut had corrugations on its butting surface with the trunnion on the camshaft lever. These fixed the clearance of the shoes from the drum in the brake-off position and also prevented the nuts from slacking off. The short brake operating rod is very close to being in line with the kingpin axis; the considerable curvature in the upper suspension levers will be noted, required to accommodate the combination of the considerable wheel movements and the 6° droop of the initially parallel levers, upper and lower, which resulted in a slightly raised front roll centre.

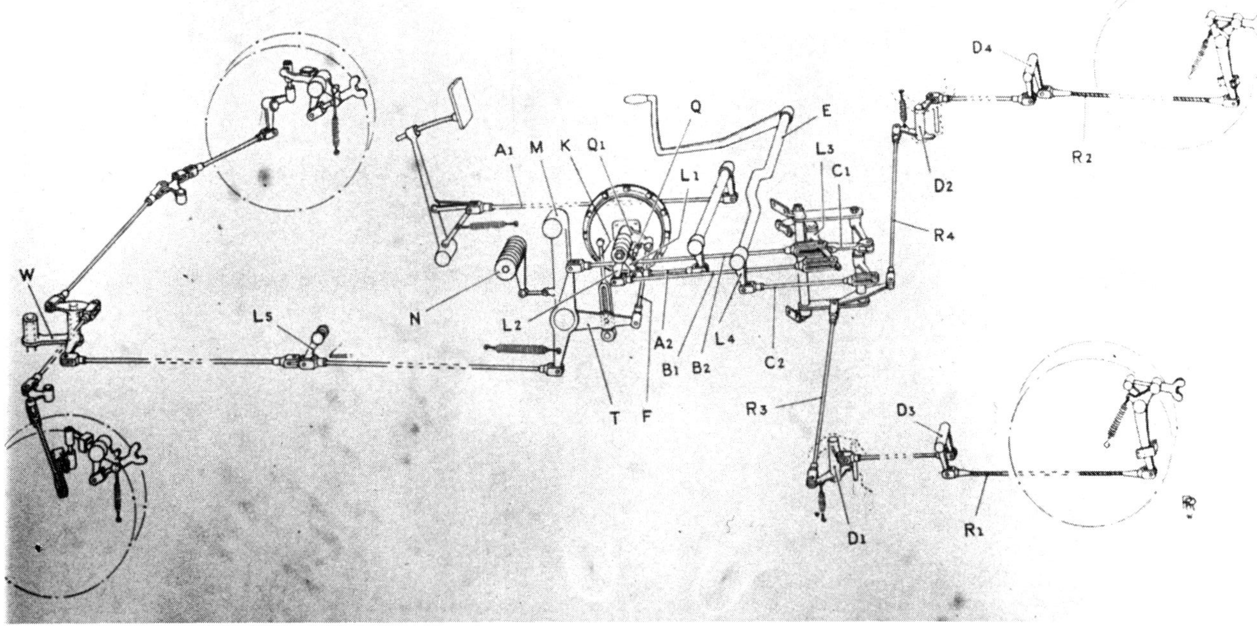

Figure 139 The complete brake linkage of the Wraith has noticeable differences from that of the original 40/50, the Phantoms I and II and the 20hp cars. The servo itself is apparently little changed and the principle of taking the direct pedal operation to the rear brake only still persists. Two major differences are the use of the bell-crank equaliser, first used on the servo, in the front and rear brake linkages for side to side balancing, in place of the bevel differentials which first appeared on the Ghost, and the combined hand and foot operation on the rear brakes to only one set of shoes in each drum. By that time this had become legal.

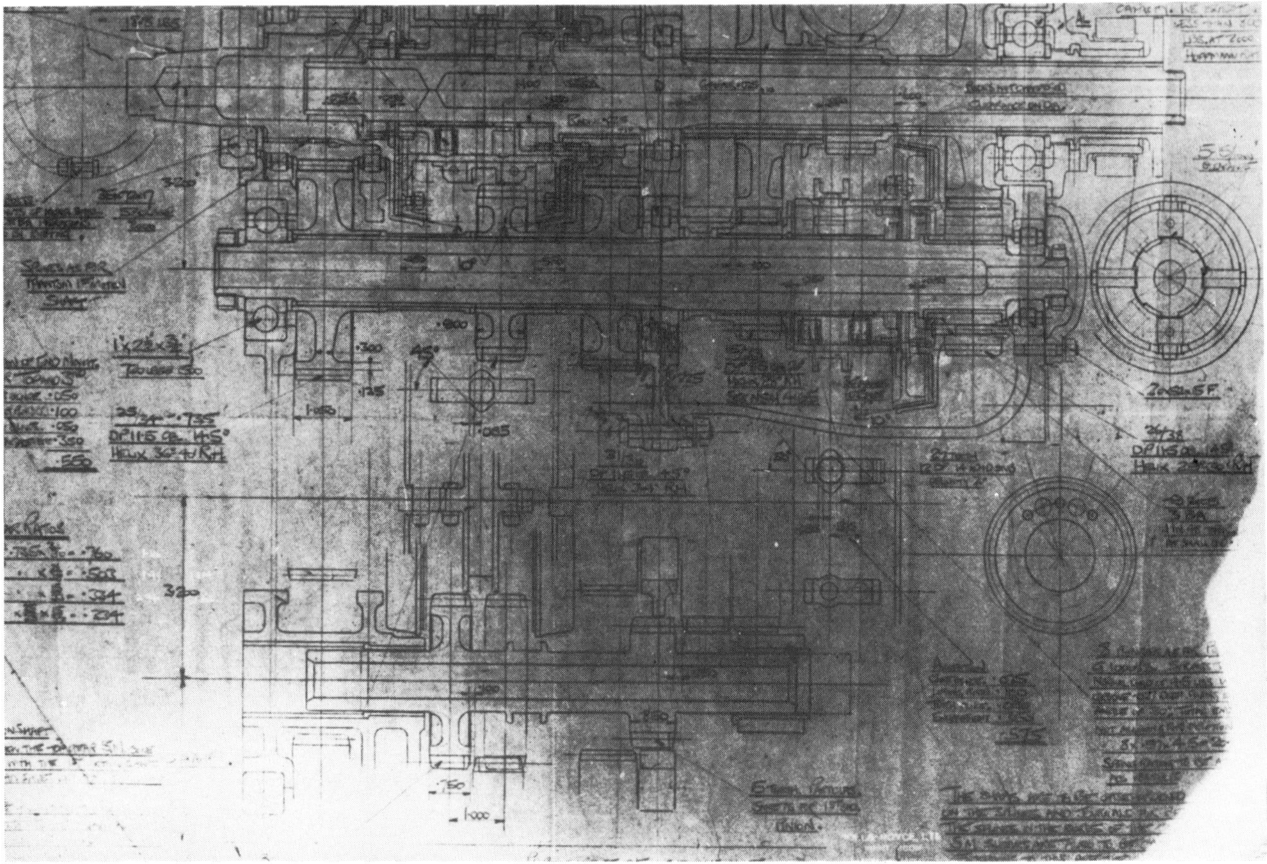

Figure 140 This scheme by Hdy of the Wraith gearbox gears and bearings is a fairly typical example of the crowded design scheme referred to in the text.

Figure 141 R.W.Harvey-Bailey - By - was an innovative designer. Although he was used as such when he first came to Rolls-Royce, his most important job over many years was as Technical Production Engineer, responsible for the production of the detail drawings from the design schemes, mainly from R's office, and the selection or confirmation of the most suitable material for each part. Shortly before the last war he was appointed Chief Engineer, Car Division, but during that war he effectively resumed his former activities on the aero engine drawings.

Figure 142 By's design for a front wheel drive taxi, 1908. Although it was able to meet the newly introduced 25' turning circle requirement, it never went into production. Its engine and epicyclic gearbox in line with it were across the front of the car.

Figure 143 A.J.Rowledge - Rg - was a Derby based aero engine designer, after his recruitment from Napier soon after the end of World War I.

Figure 144 W.A.Robotham - Rm - started at Rolls-Royce as a premium apprentice and for many years was responsible for the car experimental and development work. He set up the French testing establishment at Chateauroux. Rm is seen here on the right in this picture taken at his home in Duffield in 1936. In the centre is Hs and on the left Alex Taub of General Motors.

Figure 145 Ernest W Hives, (Hs), a photograph taken on the Austrian Alpine Trial of 1913.

Figure 146 R's original team at Le Canadel and St. Margaret's Bay consisted of these two designers, Albert G.Elliott on the left (E) and Maurice Olley (Oy).

Figure 147 E's final residence at West Wittering was Wickersfield House. Its location in relation to the village can be seen in fig 157.

Figure 148 The Schneider Trophy tea party on the lawn at Elmstead, after the 1929 event. On the extreme left is Jnr, Nurse Aubin is partially obscured by Hdy's head. Marmont's face can be seen in part beyond Hdy's. Ev faces us, next to R; on the extreme right is E, while the three quarter rear view in the foreground is of Da, stroking Rajah.

Figure 149 Bernard and Evelyn Day built this bungalow, 'Cloverfield'; like themselves it seems to be retiring. I think the tennis lawn was behind. The name 'Cloverfield' derives from the site before it was built on.

Figure 150 Hdy is standing behind the seated R at Hdy's drawing board, at one of the frequent visits to observe progress and make suggestions for improvement as the scheme progressed.

Figure 151 A group of West Wittering designers and their wives, taken at the unveiling of the plaque on the wall of the Camacha studio. From left to right, Bill Hardy, Harry Biraben, Pauline Cooke, Tony Cooke and Jane Hardy.

Figure 152 A group before departure; from the left we see Da, Ev, 'Monty' Marmont and Hdy, taken outside the Camacha studio.

Figure 153 Charles Jenner (Jnr), directly facing the camera, became E's assistant and eventually specialised in car engines, becoming the car engine chief designer. R is on the left and Hs on the right.

Figure 154 The house that Bill and Jane Hardy built in West Wittering, 'Clematis', the thatched roof is in keeping with its surroundings. Bill's younger brother Fred (FJH) claims to have put a lot of hard work into the garage.

Figure 155 In this case it is Ev who is standing at his board beyond the seated R.

Figure 156 The last day at West Wittering, 23rd May 1933. The drawing boards, plan filing cabinets and other impedimenta were being packed in the container on the lorry for transport to Derby by rail. The cars, from the left, are Tony Cooke's CMS Special Morris Minor, Ev's 20hp Rolls-Royce tourer, the author's FIAT eight and Da's Austin 12-6.

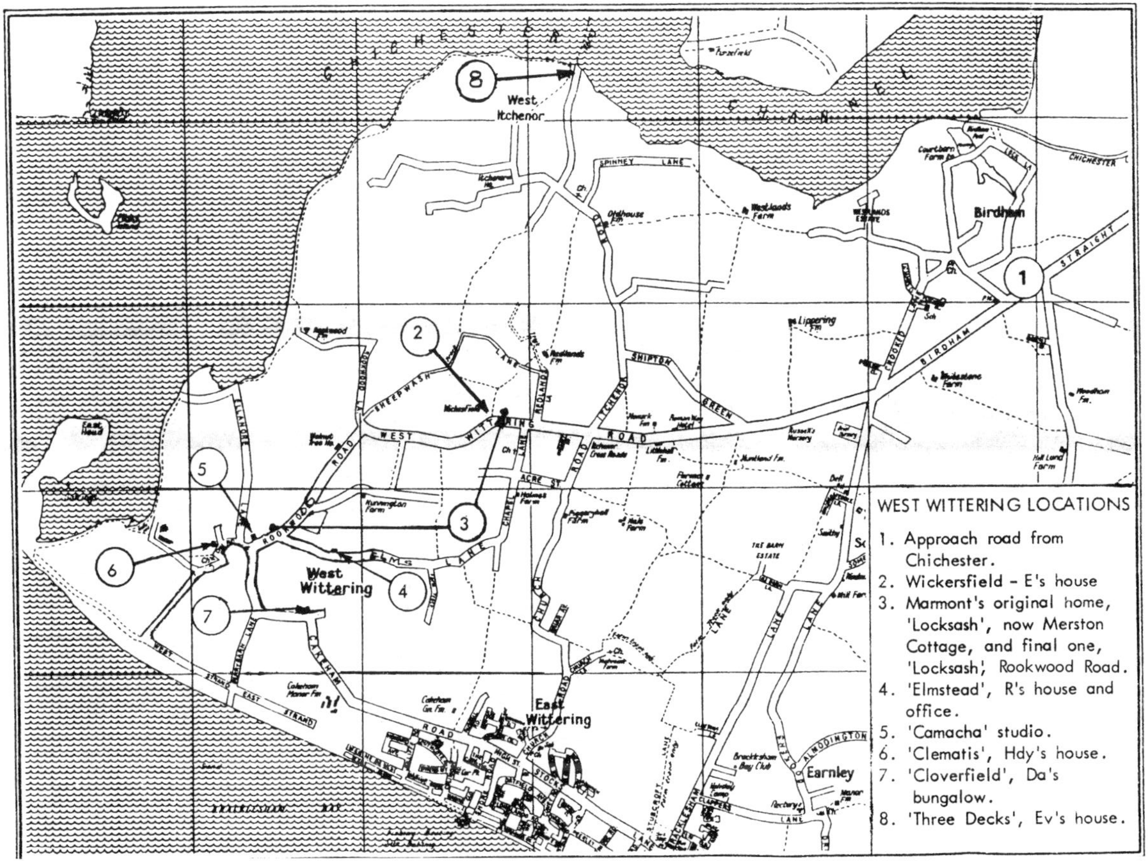

Figure 157 A plan of West Wittering and the locations at which the various people there lived. Ev and E were furthest from the job.

Figure 158 Another group outside the Camacha studio, on the day of departure: from the left, Da, Ev, Tony Cooke (AEC) and Donald Bastow (DB). They all worked in the studio.

Figure 159 R and Nurse Aubin.

Figure 160 The Rev. Arthur Tidman Gill, his wife Rose and their eleven surviving children. Max is the second from the left in the back row, Eric next but one along.

Figure 161 The unveiling of the plaque on the end wall of the studio at 'Camacha' in the village of West Wittering. The ceremony was performed by Donald J.Pepper, the vice-chairman of Rolls-Royce Limited.

Figure 162 A close-up of the plaque after the unveiling.

ROLLS-ROYCE

HERITAGE TRUST